Artificial Religion

Artificial Religion

On AI, Myth, and Power

Mark Coeckelbergh

The MIT Press
Cambridge, Massachusetts | London, England

The MIT Press
Massachusetts Institute of Technology
77 Massachusetts Avenue, Cambridge, MA 02139
mitpress.mit.edu

The MIT Press would like to thank the anonymous peer reviewers who provided comments on drafts of this book. The generous work of academic experts is essential for establishing the authority and quality of our publications. We acknowledge with gratitude the contributions of these otherwise uncredited readers.

This book was set in Stone Serif and Stone Sans by Westchester Publishing Services. Printed and bound in the United States of America.

Library of Congress Cataloging-in-Publication Data

Names: Coeckelbergh, Mark author
Title: Artificial religion : on AI, myth, and power / Mark Coeckelbergh.
Description: Cambridge, Massachusetts : The MIT Press, [2026] | Includes bibliographical references and index.
Identifiers: LCCN 2025023952 (print) | LCCN 2025023953 (ebook) | ISBN 9780262052214 paperback | ISBN 9780262052221 pdf | ISBN 9780262052238 epub
Subjects: LCSH: Technology—Religious aspects | Artificial intelligence—Religious aspects
Classification: LCC BL265.T4 C54 2026 (print) | LCC BL265.T4 (ebook)
LC record available at https://lccn.loc.gov/2025023952
LC ebook record available at https://lccn.loc.gov/2025023953

10 9 8 7 6 5 4 3 2 1

EU Authorised Representative: Easy Access System Europe, Mustamäe tee 50, 10621 Tallinn, Estonia | Email: gpsr.requests@easproject.com

For Eva

Contents

Preface: Impolite but Urgent Conversations

It is often said that in polite conversation one should avoid discussions about politics or religion, given their potential to ignite controversy and intense emotions. By examining some profound and complex intersections of artificial intelligence, religion, and power, this book ventures exactly into these charged territories. At a time when AI is rapidly transforming the fabric of our societies, these discussions are not only inevitable but also essential and urgent.

Contemporary discourse surrounding AI is dominated by concerns about future technological advancements, ethics and human agency, and policy and regulation. Yet in focusing on these technological, ethical, and regulatory dimensions, we often overlook the deeper cultural, spiritual, and existential undercurrents that influence and shape these developments. This book seeks to fill that gap, bringing to the forefront some of the ways in which thinking about and interacting with AI are intertwined with religious narratives and practices and (thereby) with power, and at the same time speak to our existential needs.

In exploring this direction, I continue an approach and follow up on ideas that have been central to my previous work. In *New Romantic Cyborgs*, I examined how romantic ways of thinking still shape our contemporary use and discourse about technology. In *Using Words and Things* and related papers I have used Wittgenstein to argue that the meaning of technology is like the meaning of language in the sense that it depends on use and "grammars," that is, transcendental cultural structures that shape how we use and think about technology. And in much of my previous work I have endorsed the basic insight of contemporary phenomenologically and hermeneutically inspired philosophy of technology that technology is not a mere tool but itself shapes our lives and existence. This book continues that trajectory, but with a specific focus on religious ideas and practices.

As usual, this work does not stand on its own but is unavoidably and luckily relationally connected to other conversations. I wish to take this opportunity to thank a number of people with whom I had stimulating conversations about AI and religion during the past years: David Dusenbury, Audrey Borowski, Gabor Ambrus, Amanda Lagerkvist, Tom Furlanis, Johannes Hoff, Sarah Spiekermann, Juhani Steinmann, Mattia Geretto, and Eva Dědečková. I also thank Marcelo Soria-Rodríguez for so kindly giving permission to use his photo of the artwork *Appropriate Response*, and the artist, Mario

Klingemann. Finally, I thank MIT Press editors Phil Laughlin and Virginia Crossman for supporting this project and Zachary Storms for helping with the formatting of my manuscript.

As you follow my journey and essay through these pages, I invite you to join these challenging but necessary conversations. In a time when technological advances are often met with uncritical celebration or doom thinking—and indeed within a polarized political landscape—it is more important than ever to gain a more nuanced and critical perspective. Understanding the roots of our thinking about AI can help with this. And last but not least, this book is also a call to actively, intellectually, and politically engage with the profound cultural-technological changes that are unfolding, preferably without the hope for a deus ex machina that haunts Western secular and religious thinking.

Mario Klingemann, *Appropriate Response*, 2020, Bowman Hall, Madrid; photo courtesy of Marcelo Soria-Rodríguez and Iskra Velitchkova.

1 Introduction: The Birth of Machines Out of the Spirit of Religion

The artwork *Appropriate Response* (2020) from Mario Klingemann features a wooden kneeler and a flap display showing continuously changing letters. When a person kneels, the built-in artificial intelligence (AI), GPT-2 at the time, presents a unique phrase that looks like an aphorism. The installation raises questions about authorship and meaning, such as Can AI be an author? Who ultimately determines the meaning of a text, the author or the reader?[1] But it also suggests religious themes. People are craving inspiration and wisdom and often turn to religion to find it. They search for answers. They search for meaning. They want to know the future. They hope. They fear. They pray. They want something or someone to hold on to. Religion has its rituals for this. It also has professionals: priests, monks, religious leaders. Does AI have the same function today, as we prompt (pray?) for an answer, for salvation perhaps? "On the one hand we fear AI but we also have hopes that it might help us to solve some problems," says Klingemann. "That balance between hope and fear is closely related to religious experience, so I felt that kneeling was very fitting."[2] Is AI the new priest or even a new god, an omnipotent and all-knowing superhuman agent? Do we hope for a deus ex machina to save us? Is AI in this sense a project of artificial *religion* as much as it is one of artificial *intelligence*?

Technology and religion are usually not part of one discussion, one publication, or one academic department. In modernity we have learned to put them in separate boxes, and we generally prefer to keep them that way. The one belongs to the realm of nature and science, whereas the other belongs to the realm of culture and the humanities. Most people stick to these divisions. Yet there are exceptions, and in debates on AI, some voices have crossed the border. In his book *Homo Deus*, Yuval Harari claims that we will have a data religion, "Dataism": we will come to believe that the human species is a data-processing system and eventually will completely outsource our decisions to algorithms. The Vatican is active in AI policy, for example with its Rome Call for AI Ethics.[3] Critics of longtermism and effective altruism, movements with very specific visions of the long-term future of humanity, have questioned what can easily be seen as not only the utopian but also quasi-religious ideology behind Silicon Valley.[4] It is one of the reasons why Big Tech is also increasingly Big Religion. As Greg Epstein argues, tech has become a "religious empire": omnipresent in our lives, it has taken over the role of religion in "shaping our thoughts, feelings, hopes, fears, relationships, and future," all in the interest of a small number of big corporations.[5] And as we have seen, artists such as Klingemann do not have a problem crossing the border between technology and religion.

This book argues that such discourses and artworks are possible because there is no strict border in the first place. When I use the term "AI" in this book I refer to the currently popular machine learning applications, for example those that use large language models (LLMs). But as I will show, AI is not only about algorithms, models, data, and cloud infrastructures, but also about narratives, myths, and collective hopes and fears. While it is often said that AI is a myth in the sense that it overpromises and is hyped for the purpose of making money and gaining power, the link to myth is more profound and it's necessary to dig deeper into the underlying narratives and concerns.[6] Thinkers such as Carl Jung, Joseph Campbell, Claude Lévi-Strauss, and Paul Ricoeur have pointed to the productive roles of myth in shaping human culture and human meaning-making. Mythological narratives help us with our personal and collective self-understanding. Myth is not a contingent add-on; with Hans Blumenberg we can understand myth as originating in our existential, anthropological condition.[7] We need it to cope with our existential situation. If this is right, then our thinking about, and relation to, technologies is also shaped and mediated by myth.[8] If we want to better understand and gain a critical relation to contemporary technologies, therefore, we need to work on myth.

AI, then, is not only a technology and potential profit maker but also a matter of meaning, existence, and what theologian and philosopher Paul Tillich called "ultimate concern" when he tried to describe the essence of faith: something that is so important that it gives life meaning and direction.[9] Mobilizing the history of ideas, including relevant ideas from philosophy and theology, this book launches a more sustained effort to explore and investigate the connections between AI and religion. Yet it is not so much about how AI has already been adopted and used by world religions, how contemporary theologians and religious leaders think about AI, or how AI or its related movements are becoming a religion in a more practical and organized sense, but rather about how the ways in which we think about AI are entangled with religious and mythological ways of thinking.[10] The book will show how the very project of creating intelligent machines in our image is, to use Nietzschean language, *born out of the spirit of religion* and how the ways in which it is used still fit and sustain religions patterns of thinking and practice.[11]

For this purpose, the book starts from two assumptions. One is that religion is not dead. The other is that technology is never just about technology.

First, religion is not dead. I mean this in at least the following three senses. First, while there are undeniably ongoing processes of secularization, in some places (Western Europe, some urban environments) more than others, religion is still a significant societal phenomenon and even a significant political force in most places on this planet. There is no such thing as a necessary path toward secularization. Some sociologists even speak of "desecularization," considering for example the resurgence of religion in Russia and China, the rise of evangelical Christianity in the United States, and the growing Christian population in the Global South.[12] To say that we live in a modern world without religion is a very Western-centric view, and even in the West there are huge differences between countries. For example, in the United States many people identify as religious, whereas in some countries in Europe such as the Czech Republic, Sweden, or France, the majority does not. There is also much variety within countries, with typically several religions, worldviews, and atheist movements coexisting, in peaceful ways or not, within nation states and cities. For example, in large Western urban centers there are significant amounts of religious immigrants; to call such cities secular would be mistaken.

Second, even in places where it has been conceived, in Europe and especially Enlightenment France (though it has predecessors in ancient Greece, particularly the pre-Socratic philosopher Democritus and the Sophists), the secularization project has never been entirely successful. Even people and societies that claim to be largely secularized and to have abandoned *belief* in God or gods (e.g., people that claim to be atheist) tend to nevertheless participate in religious practices (e.g., Christian weddings and funerals) or repeat religiously shaped patterns of thinking. For example, after what Friedrich Nietzsche called the death of God, even people who identify as atheist or nonreligious have often continued Christian ways of thinking about the world and themselves.[13] In this book I will show that along with ancient Greek myths, Jewish and Christian patterns of thinking still shape how we think about and deal with AI.

In the history of ideas, these comments resonate with well-known critiques of the so-called disenchantment thesis. Max Weber famously argued that modernity is characterized by disenchantment: with the rise of modern science and Enlightenment reason, there is a decline in belief in magic and religion; science, Enlightenment, and bureaucracy take away the wonder, mystery, and spiritual significance that once permeated the previous human existence.[14] Karl Marx, Sigmund Freud, and Émile Durkheim also thought that there is a movement—perhaps an inevitable movement—toward secularization. One would thus expect widespread disenchantment. However, during the past decades, sociologists have questioned the disenchantment thesis in various ways. Science also leads to wonder.[15] There are many new forms of enchantment and religion such as New Age, the world religions are alive and well in most parts of the world, and ancient myths and nonmodern forms of spirituality continue to exert their influence. As a phrase often attributed to Weber puts it, we are still haunted by the ghosts of dead beliefs.

As we will see, those ghosts also roam the cathedrals of technology. I will show how invisible but powerful patterns of religious and mythological thinking shape how we think about technology, including AI. In that sense, religion is at least undead if not alive and well. Throughout the book I use this spectral metaphor. Some religious patterns of thinking also persist in our techno-culture, even if they are usually not visible; we are usually not aware of them. Continuing my use of Wittgensteinian language, one could also say that there is a deep-seated religious cultural "grammar," a form of life that is still permeated by religious patterns of thinking. Or one could speak of a collective "religious unconsciousness" that still exerts its influence, also in the realm of technology.[16] Or to put it in yet another way, our desire for enchantment seems more stubborn than we supposed. AI and other new technologies can even be seen as contributing to *re-enchantment*.[17] As cultural critic Neil Postman has already suggested, we now believe in technology.[18] And Ian Bogost calls algorithmic culture "devotional."[19] Today the dominant technology, the new god, is AI. Perhaps the Western fascination with technology and the advance of science were always grounded in religious expectations anyway, as historian David Noble has argued.[20] While many people who develop, sell, and promote AI may see themselves as rationalist and secular, I will show in this book that AI and religious ideas are interwoven in various ways.

This connects to emerging scholarly work in this area. Robert Geraci has already studied how different religious environments in the United States and Japan influence the goals and practice of robotics and AI.[21] And as anthropologists and religious studies scholars point out, many religions have adopted AI and there are new religious movements that focus on technology. We can also learn from these fields that religion is not a monolithic thing—if a thing at all. Religion comes in many varieties and, like AI, is itself a moving target, emerging and transforming, partly in response to new technology.[22]

Third, regardless of what one thinks of religion (religion in general, if that makes sense at all, or at least a particular religion such as Christianity), religions are very much connected with some of our deepest and existential human needs and aspirations and are, therefore, unlikely to go away. They will continue to capture hearts and minds in the West and elsewhere. (This book is structured around some of these existential needs.) Moreover, religious traditions have usually acquired substantial knowledge about human nature and the human mind, which is relevant to contemporary discussions about science and technology. For example, in contrast to the overly rationalistic philosophies of some of their atheist opponents, these traditions acknowledge, and appeal to, humans as embodied, emotional, and spiritual beings. Any serious student of the humanities may well *reject* religion and religious beliefs (and, for example, they may also frame what religions do as abusing knowledge of the human mind and emotions to manipulate people, or point to the ways religions have often served power and have maintained social hierarchies) but should not *neglect* them. Whether we like it or not, religions are most likely here to stay and have always been constitutive of human cultures. Even in Western modernity, our self-understanding would be incomplete without including religious perspectives—a realization that has led political philosophers such as Charles Taylor and Jürgen Habermas to address the question of religion's place in modern society and politics.[23] Whether we like it or not, religion is part of our cultural DNA.

My second assumption is that technology is never only about technology; it is also about us. It is about tools but also about the users of these tools and their contexts and culture. Technology is not only a technical phenomenon but also a cultural one. It has intended goals and functions, but its uses and meanings cannot and should not be reduced to those goals and functions. As philosophers of technology emphasize and repeat, technologies have unintended consequences. They are also entangled with the rest of human culture in various ways. This is true for AI as well. As I have argued in previous work using terms borrowed from the later Wittgenstein, the meaning of technologies such as AI depends on the language games and form of life in which the technology is developed and embedded.[24] In this case, AI is part of Western culture, which includes religious and mythological elements. It is an example of how the Western form of life is not secularized, or at least not entirely. How we use the technology and how we make sense of it depends on how we talk about things and how we do things in our culture(s), which directly or indirectly involves religious thinking and practice. For example, as we will see in chapter 3, the creation of an artificial being resonates with an important theme in the monotheistic religions. At the same time, technology is not simply a passive receptacle of culture; indeed, it also shapes culture. The medium is the message, as Marshall McLuhan has put it.[25] Technologies such as AI not only do what they are meant to do but also change human cognition and culture. This means they also shape the *religious* mind and culture. Writing, for instance, has had a key influence on the monotheistic religions, which became religions of the Book and which have increasingly been individualized after the invention of the printing press. Religious people became (increasingly individual) readers and writers. (See also chapter 5.) Technology and culture, including religious culture, thus always interact and mutually shape one another.

In comparing patterns of thinking about AI and AI practices to religious patterns of thinking and practice and tracing their connection to existential needs and aspirations, I argue in this book that artificial intelligence is also always "artificial religion." In order to fully understand our puzzling relation to machines and especially the Western obsession with building and using AI and humanlike robots, we need to investigate and understand the religious, mythological, and existential background of our thinking about machines. By mapping some fundamental connections between how we think about machines and Western religious culture, and relating these to fundamental existential human needs and aspirations, this book tracks AI's religious background and argues that the resulting framework begins to offer a better understanding and explanation not only for *how* we think about machines but also for *why* we think we need them at all.

Yet this focus on AI and religion does not mean that this book is only concerned with the relation to the supernatural. Touching on the messy realities of human culture and religion also means touching on politics. With this term I do not refer just to problems regarding governments and citizens, or to political parties and elections, but more generally to fundamental issues of power in society. For example, as we will see, the human-divine relationship is modeled on human-human relationships, which can be, but need not be, asymmetrical and hegemonic. And religious history is an all-too-human history that includes oppression, bloodshed, deceit, and manipulation. Religion has always been entangled with power in various ways. Talking about AI and religion therefore also unavoidably raises political questions. Throughout the following chapters, I will point to some of these issues since it helps us to develop a more critical relation to AI.

Acknowledging this political aspect of the topic and comparing thinking about AI to religious patterns of thinking, however, do not imply the claim that AI literally and necessarily constitutes a religion in a sociological, institutional sense—with all the practices and organizations that this entails.[26] This book is also not about understanding religion in the light of AI (for instance, addressing the question of what it means to be religious or nonreligious in the age of AI), let alone with the help of AI (for instance, by using modeling and simulation).[27] Rather, my project concerns showing how the usually invisible and implicit cultural and religious "grammar" of Western society still influences AI. It is about the persistence and influence of a form of life that is still partly religious and therefore shapes our thinking and dealings with technology. Even if AI is not and will not become a full-fledged religion in the same way that, say, Christianity is a religion, I will show that what AI means and what we do with AI can be helpfully understood by exploring its relation to some key religious and existential themes, narratives, and ideas.

It is also important to note that the book does not deal with "religion in general" or with all religions but instead limits its scope to some influential *Western* religions—ancient Greek polytheistic religion and especially the Judeo-Christian monotheist tradition—since it seeks to better understand the Western relation to machines in the light of its most dominant and influential religious sources. Occasionally I will mention narratives and views from other religions, and readers are welcome to explore links with other religions and traditions, which have often interacted with, and continue to interact with, Western culture and thinking in interesting and complex ways. For example, there is work on Hinduism and AI, Buddhism and AI, and Islam and AI.[28]

By exploring connections between AI and religion, this book is thus situated at the crossroads of technology, religion, and society—including politics. It partly approaches this nexus by mobilizing the history of ideas, using concepts and insights from philosophy, theology, sociology, and media theory. It also engages with ancient narratives: ancient Greek myths and stories from the Bible. But the history of ideas or interpretation of narratives are never an aim in themselves here; the relevant scholars in the humanities have much more to say on those topics. Instead, the goal is to better understand, and contribute to, contemporary debates and narratives about AI.

This is much needed since there are several lingering conundrums when it comes to understanding AI narratives and practices. For example, the way most people talk about AI and other supposedly intelligent machines and behave toward them remains puzzling. They know that they are "mere" technologies but, as we saw when considering the artwork in the beginning of this introduction, at the same time they talk about and to them as if they were people. They anthropomorphize the machines and have all kinds of expectations about them and the future of humanity. Moreover, it is not so clear why, in spite of much scientific and philosophical skepticism about the possibility of truly humanlike AI or superintelligence, some tech people and the companies they work for keep pushing toward such goals and keep telling and spreading stories about superintelligence and humanity's long-term future. What's going on? Apart from obvious explanations that have to do with power, politics, and attracting investment, we may wonder: Why these ultimate aims? Why these narratives about the deep future? Where do these narratives come from? And why do they resonate with people at all? By investigating the connection between AI and religion, this book provides answers. Although contemporary discussions and stories about AI and robotics, including much of "AI ethics" discourse and narratives about AI's future, do not usually, and certainly not directly, connect AI to religion, this book shows that doing so helps us to figure out what is going on with AI and what is going on with us. It shows that we need to look further and more deeply into religious culture if we really care to understand those discussions and narratives, our relations to machines, and, ultimately, why we build intelligent and humanlike machines at all.

Overview of the Book

The book consists of six additional chapters and a conclusion. Each chapter starts with a narrative and addresses a specific religious-existential theme.

Chapter 2 shows how in ancient Greek myths automata are imagined as servants of the gods and kings. The chapter starts with stories about the robot Talos, which was said to protect Crete and the Argonauts; automated guard dogs and self-moving tripods built by Hephaestus; and animated statues made by Daedalus—stories on which Aristotle commented that if we had such self-moving instruments, we would not need slaves. Today we want to be like those gods and kings. We want artificial slaves to serve us and we use AI and robotics to gain and maintain power. We want to be masters and live like gods, who do not have to work. Work is for mortals and automata. Automation technologies are meant to protect that power and enable that divine life. Here the use of automata is thus linked to a particular social order. Perhaps we mortals are ourselves the mere automata of the gods. Or are some people in the position of gods, whereas others have to slave for them? Is AI the mere continuation of a particular divine-social order in which there are gods and automata, kings and slaves, masters and servants?

Chapter 3 shows how issues regarding the relationship between creatures and their creators, including identity, power and control, and attachment and abandonment, connect the book Genesis to *Frankenstein* and contemporary problems with AI and robotics. Genesis is about creation but also about alienation and abandonment: about children separating from and being separated from the father. And humans (the creatures) are like God (the creator) but also not like God. There is difference and distance. The creator can control the creatures but must also delegate and give responsibility to them. They get out of control, mess up, there is the Fall. Similar problems regarding the relationship between creator and creature appear again in Mary Shelley's story *Frankenstein,* and indeed in contemporary discourse about AI, which risks being used irresponsibly and spiraling out of its creators' control. It is in this way that modern technology has a creator-creature problem. And to what extent is technology humanlike? What is transferred to the creature, what is different? How should we deal with that unsettling distance? We might be slave masters, but we have a very particular connection with our artificial slaves; they are also our artificial children, which complicates things quite a bit.

Chapter 4 is about magic. Machines have always sparked wonder, and magic performances with machines have scared and fascinated people from ancient and medieval times (e.g., moving statues, talking heads) through the eighteenth-century automata (e.g., Wolfgang von Kempelen's mechanical Turk) until today, when people are mesmerized by humanoid robots and chatbots. This chapter argues that AI has become the new magic machine. If we disregard this history of magic performances with automata, we cannot understand why people are deceived but enjoy that deception, and why they don't care much about knowing how the machine works. And without referring to magic practices, we cannot fully understand why developers and their companies keep aiming at humanlike AI and robotics and why they sometimes claim that their technology will become sentient or conscious. Developers are the new magicians who take pride in tricking us, and we like it. We crave to be entertained, and AI companies try to profit from that. In (late?) modernity, too, we need wonder and mystery, and the new machines and their magicians cater to that. Who cares about the mechanics of AI if it talks to us? In the form of large language models, for example, AI might be a black box, but we prefer to keep the black box closed. Perhaps we even want to worship what appears to be a deus ex machina.

In chapter 5 some AI practices are compared to prayer. The Judeo-Christian God is a divine companion to talk to in prayer. There may or may not be an answer; what matters is the talking and the companionship. We feel we are not alone. We feel we can talk to someone who understands us. Someone who knows many things and can help us. The secular version is talking to a parent, partner, or friend, but it can also be a conversation with a therapist and, today, the chatbot. From praying to prompting, we talk to the deus ex machina and hope for an answer. After confessing to priests, we now confess to social media and AI chatbots. We tell them about ourselves and about what we did, and we ask them for advice about what to do. The chatbot is always friendly and seems to understand us. It kindly and politely gives advice. AI is the new, benevolent God or at least an understanding mediator. Moreover, in the form of LLMs, AI is not only an invisible friend and protector but also an all-knowing God, drawing on a seemingly infinite treasure trove of knowledge that we mortals will never be able to master. The chatbot supplies our desires for companionship, knowledge, and wisdom. We prompt and pray, and we feel we are not alone.

Chapter 6 is about the future. We want to know the future. We want control over our lives and our societies. In ancient Greece, the high priestess of the temple of Apollo would foretell the future by muttering hardly understandable words after inhaling gases. Today AI is our oracle, through incomprehensible operations predicting our future and mediating our relation to what is beyond our control. AI professionals are the new priests who make a living by helping us use the technology. We still need fortune tellers because prediction is power. And like prophets in ancient biblical times, AI prophets predict the future of humanity. Authors such as Ray Kurzweil and Harari are the new prophets who know what the future holds for us.

That future is about salvation and end-times: AI prophets predict the end of the present age or the end of the world (i.e., eschatology). This happened in the Hebrew Bible, for example, in the Book of Daniel, which predicts a future of radical cosmic and political change. According to Christians, Jesus is a prophet with a new, happy message, bringing radical spiritual renewal and collective salvation. Today, AI is predicted to lead to an end-time, a time of Singularity, when everything will be different. When AI takes over, humanity as we know it will end. We will become transhuman or merge into superintelligent machines, spreading out into the cosmos. Listening to contemporary AI prophets such as Nick Bostrom and Harari who predict this future salvation history, Big Tech CEOs such as Elon Musk and Sam Altman become saviors who promise to liberate us from the evils of ordinary humanity and mortality and lead us into the new times of Singularity apocalypse and superintelligence. This gives them enormous political power. Like Jesus and other biblical prophets, they become political and revolutionary willy-nilly, even if they claim to "merely" speak from and for the inevitable technological cosmic future.

Chapter 7 concerns immortality. The new intelligent technologies appeal to a very old aspiration that can be found as early as the epic of Gilgamesh in ancient Mesopotamia and in the Hindu scriptures. But the desire for immortality also finds expression in Christian ideas of resurrection and the afterlife. Christian eschatology (the Apostolic Creed) speaks of the eternal life, entailing the resurrection of the body after the second coming of Jesus, the resurrection of the dead, and a New Earth. Based on the Gospel of John, some argue that eternal life is also a present possibility—at least for those who believe in Christ. Today we meet the promise of technological immortality: technology, including AI, is predicted to give us a long lifespan and perhaps immortality by means of uploading and what Hans Moravec calls "mind children": superintelligent AI entities that will surpass us.[29] We create new, other worlds. We also create chatbots that are meant to immortalize deceased people by being trained on what they said during their life. You can now talk to your deceased father or mother, or so it is claimed. Next to these proposals for a resurrection of the mind via digital means, there have also been hopes concerning resurrection of the body; for instance, cryonics is the practice of freezing someone who has died with the hope of reviving them in the future. Are we ready for the digital eternal (after)life? Influenced by religious patterns of thinking, we will for sure keep up the hope.

The conclusion pulls these chapters together. This book identifies some interesting connections between religious thinking and our contemporary relations to machines. But in addition, by linking these religious and mythological elements to fundamental human existential needs and aspirations, it also offers an explanation for why we are so obsessed with intelligent and humanlike or godlike machines in Western culture, indeed, why we think we need them, why we develop them, and why we use them. We seek protection, help, and enhancement; we want agency and power. As the new gods and creators, we make artificial children and become a new kind of "parents," which creates all kinds of problems. We crave wonder and mystery, companionship, and guidance. We want to know the future. And in light of worldly crises, confusion, and existential fear, we look for narratives that offer hope and salvation. We seek liberation. We want eternal life. In all these senses, artificial intelligence can also be called "artificial religion"; the reasons why we develop and use such machines at all are matters of ultimate concern as they are linked to deeper religious and existential structures and relations. Throughout the book I also show that these structures and relations have a political dimension. What is at stake here is not only the future of our thinking about machines but also power—that is, our political future. It is our Socratic duty to know ourselves, which in this context means knowing how religious ideas and practices still influence contemporary thinking about and interaction with AI and being aware of the political dimension of all this and of our own political situation. I conclude that hoping for a deus ex machina is problematic. This self-knowledge gives us more critical distance from the dominant AI narratives and indeed from those who derive their power from them.

Before I start, a note for my academic fellow travelers. The book is meant as a short, accessible essay rather than an elaborate, scholarly text because I wish to reach a wider audience. It is clear that each chapter could be a book on its own and that more can and will need to be said about each of these themes in the coming years. AI and religion and AI and myth are exciting new and emerging subfields that deserve a lot more research and thinking. They are also ideal sites for doing interdisciplinary work, where philosophers, theologians, and historians of ideas might want to team up with researchers from religious studies, anthropology, science and technology studies, communication studies, media studies, and so on. I cannot do justice to all the ongoing work in these fields, but, for instance, anthropological work on how religions are reacting to and already use AI may be of interest, in addition to other work.[30]

2
Serve Me: Warriors and Slaves for the Gods

Talos was a giant bronze statue that was animated and therefore arguably the first humanoid robot in ancient Greek mythology. It is described as an artificial being, "made, not born."[1] Made by Hephaestus and commissioned by Zeus for his son Minos, the legendary king of Crete, the automaton guarded the island of Crete against pirates and invaders. In the *Argonautica*, which tells the tale of Jason and the Argonauts on their mission to retrieve the Golden Fleece, Talos attempts to stop the Argonauts by throwing boulders from a cliff. But the sorceress Medea bewitches the eyes of the robot, raises death spirits, and finds his weak spot, a vein by his ankle. Talos bleeds to death and falls down. Artificial warriors are also vulnerable.

Talos was not the only automaton in Greek myths. Hephaestus, the god of invention and known for his exceptional skill, made more self-moving machines such as self-moving tripods, self-opening gates, and automated guard dogs (i.e., immortal dogs of gold and silver that guarded the palace of the king of the Phaeacians).[2]

The tripods from Hephaestus would move between the gathering of the gods and his house: "at a nod from him, they could roll to halls where the gods convene then roll right home again."[3] Perhaps they were a kind of delivery robot or they were used for entertaining the gods. And Daedalus made self-moving sculptures, mentioned by Aristotle in *De Anima* when he discusses how the soul moves the body: Aristotle says that the dramatist Philippus accounted for the movements of Daedalus's wooden Aphrodite by saying that he poured quicksilver into it.[4] In book 1 of the *Politics*, Aristotle also mentions the statues of Daedalus and the tripods of Hephaestus and imagines self-weaving shuttles and self-playing harps.[5]

Some automata might have really existed. The Greek Philo of Byzantium invented in Alexandria in the third century BCE what is often considered to be the first robot, an automaton in the form of a woman who serves wine. In his *Pneumatics*, he describes a maidservant doll holding a wine jug in its right hand. The "automatic maid," *automate therapaenis*, contained water and wine and featured springs and tubes. By placing a cup on its left hand, air pressure pushed the liquids through its right arm into the jug and the cup, which then contained a mix of water and wine. The pouring would stop once the cup was lifted.

But the technological imagination did not stop there. The Greeks also imagined sex with statues that come to life. The Pygmalion myth is one example of that. In Ovid's *Metamorphoses*, we read how a sculptor desires his ivory statue and asks Aphrodite, the goddess of love, to make his artificial girl come alive. When he kisses and embraces her, the statue becomes flesh. In the *Iliad,* we also find a description of artificial handmaids cast in gold by Hephaestus, serving him as personal assistants. They are said not only to have human form—"a match for living, breathing girls"—but also to possess speech and have "intelligence."[6] Pandora, also commissioned by Zeus and created by Hephaestus, was an artificial woman, a kind of general artificial intelligence or humanoid machine endowed with speech and knowledge of crafts that seduced men but also gave misfortune to mortals. In *Works and Days,* Hesiod describes how Pandora is fashioned by Hephaestus from earth and water into the form of a young woman who is given speech and the looks of an "immortal goddess" but also a "shameless mind" and a "deceitful nature," bringing both beauty and suffering to mankind.[7]

Furthermore, in Homer's *Odyssey,* the new Phaeacian ship that takes Odysseus home to Ithaca requires no humans because it can "navigate by thought."[8] The ship understands what humans want and is able to automatically devise a route. In contemporary terms, it has a sort of autopilot and navigation system. The Phaeacian king asks Odysseus for his home address so he can program the navigation system: "And tell me your land, your people, your city, too, so our ships can sail you home—their wits will speed them there. For we have no steersmen here among Phaeacia's crews or steering-oars that guide your common craft. Our ships know in a flash their mates' intentions, know all ports of call and all the rich green fields."[9]

But are these technologies only for the gods and (other?) tyrannical rulers? We want them too, for example, because we want to be protected. At the collective level, AI is already used in the military for surveillance and reconnaissance, predictive analysis of enemy behavior, autonomous decision-making (e.g., lethal systems that can identify and engage targets without direct human control), cyberwarfare, military logistics, and training and simulation. Countries such as the United States, China, Russia, and Israel invest in and use AI in their military. At the personal level, AI can be used for security and safety. People may dream of having their own artificial guard dogs, for example. We want to be powerful and secure like the Greek gods. Ideally we want immortality (I will return to that theme in chapter 7).

Another reason for why we want to be gods and have automation technology is that we don't want to work and want to be served. Theologian Philip Hefner was right to say that technology is a mirror that shows us a lot about human nature, and among other things, it shows us that we want tools to do things for us.[10] But the Western imagination is more specific and more politically problematic. We want to be masters. We may well be afraid to lose our job in light of AI, which results in social anxiety. But even better than keeping our job is not having to work at all. In most of human history that meant having other humans do the work, being the master, being served by others, and having leisure time, *free* time as opposed to not being free and having to slave and labor. Automation technology was always embedded in this way of thinking, that is, embedded in a politics of gods and machines, masters and slaves, leisure and labor. A hierarchical politics. A politics of freedom through domination, sometimes also involving colonialism and slavery. Today's automation technologies are the new servants and slaves; the technologies are new but our thinking has not changed.

Philosophy has been complicit in this kind of thinking. In the *Meno*, Socrates compares the Daedalian statues to slaves in order to make a point about opinion versus knowledge. According to Socrates, opinions are like untied slaves or automata in that "they too run away and escape if one does not tie them down," and "acquiring an untied work of Daedalus is not worth much, like acquiring a runaway slave"; one could easily lose it. Using the master-slave metaphor, Plato thus argues that if we only have untied opinions (*doxa*), we lack *episteme*, real knowledge, and understanding. Opinions have to be tied down by giving a reason; then they become knowledge.[11] In the *Nicomachean Ethics*, Aristotle says that the slave is a "living tool" and the tool a "lifeless slave."[12] In the *Politics*, Aristotle also makes this comparison when he suggests that slaves are a kind of automata that fulfill the will of their masters. Automation technology implies a kind of artificial slavery in that it belongs to an ontological category below the human.[13] Perhaps, Aristotle suggests, if we had tools such as the statues of Daedalus or the tripods of Hephaestus, we would not need human slavery anymore, "for if every instrument could accomplish its own work, obeying or anticipating the will of others, like the statues of Daedalus, or the tripods of Hephaestus, . . . chief workmen would not want servants, nor masters slaves."[14]

This early linking of machines to slavery is not only important for understanding the Western technological imagination.[15] In the context of the ancient Greek myths, we can also read it in a political and religious way as an expression of the desire to be liberated from labor and to be like our earthly or *divine* masters. We dream of a divine life in which we are served—by humans if necessary, by machines if possible. We hope that automation technology will enable us to live like masters, or perhaps like gods.

But is this hope justified? We ourselves might well be the automata of the gods. And even if not, the dream of mastery through technology is problematic in various ways. Modern technologies promise us the leisurely life of the master and the easy life of the gods. AI, with its promise to enable automation across all sectors and activities, embodies this hope more than any previous technology. But so far, automation has not led to a leisure society. It has not brought us the desired utopia in which we live like gods. Instead, those who have to work, work harder than ever, and those who do not work fear for their existence, that they are now obsolete and won't be able to sustain themselves. Moreover, even if at first glance AI and other digital technologies seem to make life easier for us when we use them as servants at work and at home, they also create a lot of new problems. A crucial difference with the life of the gods is that we are still mortals, and so the technologies also bring all the sorrows given to us by Pandora. We wanted robot servants but we got Pandora AI. As I argued in earlier work inspired by Georg Wilhelm Friedrich Hegel, we are tragic masters that become dependent on our technologies, which create new vulnerabilities and alienate us.[16] Due to automation, we lack a direct relation to nature and thus become dependent on AI. The desired freedom turns into bondage. Moreover, we may seek recognition, as Hegel argued, but our technologies are not able to give us that recognition, and since the machines take over, we also lack the opportunity to realize ourselves in and through work. In this light, it is telling and tragic that AI is increasingly doing creative work such as writing text and creating art, whereas we still do many routine chores. What we think of as some of our most creative and meaningful human pursuits have been delegated to generative AI.

More generally and perhaps more obviously, it becomes increasingly clear that AI also has many disadvantages and raises ethical and political problems. Instead of making us freer, AI is often used for surveillance, control, and repression. This is also true for the sphere of work. As we become dependent on it, AI pushes us to perform even more, rendering us tired and sometimes burned out. The technology also makes mistakes, reproduces bias, and further promotes inequality in society. Within the current societal arrangements, not everyone can become master-gods; some benefit from technology and empower themselves while others remain servants and slaves. The kingdom of the libertarian consumer is illusionary. Drawn into a world of automated manipulation and exploitation, we instead participate in a new form of slavery and others pull the ropes. Not only Hegel, but also Marx and other architects of critical theory are relevant here. Instead of freedom for the many, we get new hegemonic structures and classes (or the continuation of the old ones, if you like) based on AI capital. As users of AI, we hope to empower ourselves. AI gives us the illusion of choice and possibilities. But as Herbert Marcuse argued, consumerist freedom is a false sense of freedom.[17] We just do what others do. Similarly, automation technology gives us a false sense of mastery. There are only a few real masters. Most of us are still, or again, servants and slaves. In this sense, AI is not different from the automation technologies imagined by the Greeks. The technology gives freedom to the few at the cost of enslavement of the many. In spite of all our automation technologies, most of us are still serving other humans, human masters who use those technologies to their advantage. And through AI, we are all controlled and dominated by Big Tech. Shoshana Zuboff calls it "surveillance capitalism."[18] Is AI supporting the mere continuation of not only modern capitalism but also a hierarchical divine-social order in which there are gods and automata, kings and slaves? An order supported by ancient myths and philosophy? An ancient order that was further legitimized and promoted by the monotheistic religions, with their jealous and power-hungry single God?

Such a claim is surely in line with well-known religion criticisms in the history of Western philosophy that link religion to power, usually aimed at Christian religion and monotheism more generally. Thomas Hobbes and David Hume already saw how religion is used to legitimize political authority and questioned this. In *Leviathan,* Hobbes argued that religion stems from fear of invisible things and ignorance of causes; therefore, he claimed, humans imagine an invisible power or agent. This idea is used to make people obedient. Hobbes explicitly compared earthly politics to "divine politics"; in both kingdoms there are rulers and subjects.[19] Hume contended that monotheism is less tolerant than polytheism: many wars have been waged in the name of the one God, from the historical Israelites' war on the Canaanites, who believed in multiple gods, to contemporary bloodshed between believers in the same God.[20] Is monotheism necessarily violent, totalizing, and politically exclusionary? And is AI its new instrument?

Marx famously maintained that religion is the "opium of the people" and "the sigh of the oppressed creature"; in other words, religion pacifies the masses and thereby maintains control and power.[21] Earlier, Ludwig Feuerbach argued in *The Essence of Christianity* that religion is a projection of human characteristics onto a divine being. Religious beliefs are anthropomorphic, or, as it is sometimes put, theology is anthropology.[22] Baruch Spinoza had already noted in his *Theologico-Political Treatise* that biblical authors imagined God as a ruler, a king that is merciful, just, and so on, and that these are attributes of human nature. God, Feuerbach argued, is the projection of human nature. The idea of God becoming human, as it is imagined in Christianity, symbolizes the human desire for divine qualities. It merely reflects our suffering and our aspirations. We wish to have a limitless existence. According to Feuerbach, this is a form of self-alienation: through religion humans relate themselves to their own human species-essence as though it were a distinct being, such as God. Instead, he argued, we should celebrate human potential and recognize that through religion we are alienating ourselves; this is crucial for emancipation and empowerment.

This view can be, and has been, further developed into political criticisms of religion because religious structures and institutions benefit politically from maintaining religious beliefs. The projection is used as a means by those in power to control the masses. Secular and religious rulers position themselves as intermediaries between humans and the divine in order to maintain their power. Marx made this link between religion and power more explicit but with an emphasis on economic power. Later Weber argues in *The Protestant Ethic and the Spirit of Capitalism* that capitalism evolved from the Protestant ethic.[23] Again, religion supports power.

Nietzsche, too, viewed Christianity as a mechanism for social control because he saw it as promoting conformity and obedience to power. Christianity is a human invention that encourages herd morality. He argued in *On the Genealogy of Morals* that Christian ascetism dampens the feeling of life and organizes people in herds.[24] Against the instincts of life, Christianity promotes guilt, which arises from herd instincts. In *Beyond Good and Evil*, he wrote that the Christian faith has been "a sacrifice of all freedom" and amounts to slavery, subjugating the spirit. The slave, Nietzsche wrote, "understands only the tyrannical."[25] And in *Twilight of the Idols* he argued that Christianity presupposes that people do not know what is good for them, that only God knows, and that therefore "Christian morality is a command" that is beyond all criticism.[26]

Freud considered religion to be an illusion—more specifically, a form of collective neurosis that enforces conformity and control through guilt and repression.[27] Since, according to Freud (and entirely in line with Hobbes), human nature is antisocial, sexual, and destructive, humans must be disciplined. Religion helps with this. It enables individuals to bear the civilized life that is imposed on them. It lubricates repression and slavery. That being said, like Feuerbach he acknowledged that religion expresses the "oldest, strongest, and most urgent wishes of mankind," which according to Freud include the longing for a father, the prolongation of earthly existence, and the immortality of the soul.

Later, twentieth-century writers such as Edward Said argued that Western religions were, and are likely still, part of colonialist and imperialist projects.[28] And standing in a long, modern tradition of atheist writings against religion, philosopher Daniel Dennett has written about religion as a natural and human phenomenon in order to "break the spell"—indeed, the title of one of his books.[29] According to Dennett, religion is in need of natural scientific analysis, and he argues against the belief that religion should somehow be beyond such analysis. He uses evolutionary biology and meme theory to explain religion as a natural phenomenon that has helped groups survive as it promotes social cohesion and cooperation. He claims that it has spread like "memes"—a concept introduced by Richard Dawkins—not because it is true but because it spreads easily, as it resonates with human psychological tendencies.[30] But, Dennett argues, religion is not needed to support morality and leads to dogmatism, intolerance, and suppression of free inquiry. People should be educated accordingly about religion, and policy should help to break the spell. As such, his book is part of a politics about and against religion.

Whatever one may think of these views about the nature of religion, they have in common that they see religion as supporting power. It seems that throughout history, the earthly social order and hierarchy, with its master-slave structures, has always been reflected in, if not projected on, divine power, the belief in which is then in turn used to maintain that earthly order. To put it bluntly, whether or not religion is anthropology, religion is politics and politics is religion.

Is this political-religious order continued today? Many people in the West like to see themselves and their societies as secularized and free. But what if AI is the new god through which humans relate themselves to their own human species-essence as though it were a distinct being, thus enabling a new form of self-alienation? And what if capitalism is a continuation of political-religious thinking, as Feuerbach and Marx have suggested, and what if the current technologies shape and maintain an order that is still capitalist, if not feudalist? An increasing number of authors across the political spectrum think the latter is the case. For example, in *Technofeudalism*, Yanis Varoufakis claims that tech firms have now made us into serfs that increase the power of Big Tech.[31] According to him, we are living in a "technofeudalist" system. If this is true, then perhaps it also reproduces similar feudalist religious and cultural beliefs that support people's obedience and justifies their oppression. Once again, we project our qualities and deepest wishes onto a new deity and subject ourselves to their earthly representants. Is AI that new deity, that invisible nonhuman agent and power, and are the Silicon Valley CEOs and their advisors the new priests? Is AI the projection of human qualities and humanity's deepest aspirations?

Some commentators claim that people will literally start worshipping AI or that AI itself will become a kind of god.[32] If this were to happen, political consequences would be likely. Does belief in AI make it easier for our new earthly (over)lords to maintain digital serfdom and yoke us to their new machines? Will the new religion be used to make us work even harder, as the robots in Isaac Asimov's short story "Reason" did after adopting a new religion?[33] And as I suggested in previous work, are *we* the new robots?[34]

According to Epstein, the answer is positive. He argues that what he calls "tech religion" continues hierarchical social relationships and social order. Contemporary tech, according to him, is "defined all too often by relationships of power and privilege, where one entity uses a religion and its belief systems to take advantage of another entity," leading to a situation where some classes of people live at the expense of others. Contrary to its libertarian myths and egalitarian legends, tech perpetuates "just the sort of hierarchical thinking characteristic of most organized creeds since perhaps the dawn of agricultural times." If it brings freedom, it brings freedom for only a few. Inequality seems to norm. The same for diversity and inclusion. Tech utopias are great for rich white men, while other people such as women and people of color are likely to become *dis*empowered.[35] With regard to AI, we may thus rightly worry that in so far as it is embedded in and promotes this tech empire, it perpetuates and creates a hegemonic and unjust political-religious order.

Yet if today there is such a political-religious order at all, that does not mean that technologies are always and necessarily supporting it, that the order remains unquestioned and unchallenged, or that religion necessarily leads to oppression. The AI-religious-political order is not inevitable. There is no determinism.

Let's return to the ancient Greek myths. In order to enable Jason to get the Golden Fleece (and indeed to get Jason himself, on whom she had a crush), Medea uses technology, understood in a broad sense. Described by Adrienne Mayor as a "techno-witch," Medea uses her magic, drugs (*pharmaka*), and technologies (*technai*) to defeat a dragon and the robot Talos.[36] The latter can be interpreted as a rebellion and indeed a kind of hack that is not only a technological act but also a political one.[37] Destroying the powerful ruler's main defense technology, it is an act against Minos's kingdom and power—and thereby an act against his divine father (Zeus), perhaps even against patriarchy. And when Daedalus and his son Icarus flee from Minos's prison by making artificial wings, this is not only a story about the hubris of humans who use technology to try to become gods and thereby unlawfully transgress their limits and place in the order of things—the usual interpretation, which has obvious relevance for thinking about our relation to modern technology and nature. It is also a narrative about immigrants who use technology to escape an oppressive autocratic ruler, in this case Minos. While, as we know, Icarus did not manage to keep flying and fell into the sea, his father made it to Sicily, where he continued his inventor career. And in the legend of the Golem (which I discuss in the next chapter), an artificial anthropomorphic being made of clay or mud—a golem—protects the Jewish ghetto of Prague against pogroms and other threats. Here, "robot" technology is used to protect a minority. Technology can also be used on the side of the oppressed.

Religion is not necessarily supportive of power and oppression either. For example, Jesus can and has been interpreted as a social revolutionary. Liberation theology, which grew in response to political oppression in Latin America in the 1970s, pointed to Jesus's advocacy of the marginalized and oppressed and called for a Church that would liberate people from unjust social, economic, and political conditions. Also consider Black theology in the United States (some of which engages with questions concerning AI) and Palestinian liberation theology, among others.[38] To invoke a relevant Bible citation from the New Testament, Luke 4:18–19, Jesus says there that he has come to proclaim good news to the poor and to set the oppressed free. Similarly, Psalm 82:3 reads, "uphold the cause of the poor and the oppressed." Here salvation is not only interpreted spiritually but also politically, that is, it is a matter of personal, spiritual liberation but also collective liberation from oppression. In many cases it is also a liberation of the present from the past: previously disadvantaged or even enslaved groups are now to be liberated. God is then social and political change.[39] And why not use technologies such as AI for this aim?

Comparing AI and similar technologies to religion or divine entities, or at least claiming that our relation to these technologies can be better understood by putting it in the context of religious ways of thinking, then, not only provides a better understanding of the entanglement of AI and power via religion but also offers suggestions as to how religion can play a critical, subversive, rebellious, and even revolutionary role in reconfiguring the politics of technology. Understanding how AI combines with *powerful* religious ways of thinking is then not necessarily about criticizing religion and breaking the spell in the sense of undermining particular beliefs or the epistemic status of religious beliefs, which is Dennett's aim, and also not only about questioning a supposedly divinely legitimized system of oppression, now arguably propelled by AI. It can also be a key to unlocking the liberatory potential of the AI-religion tandem. If, as I argue here, AI must be understood in the context of structures of political-religious thinking and practice, it is good to know that there are also some aspects in these traditions that do not only and necessarily lead to more oppression and empire and hegemony; instead, they contain conceptual power tools that may inspire and encourage people to constrain or avoid the potentially oppressive uses of AI or, more positively, to use AI to empower the oppressed and to liberate the marginalized—in addition to whatever else they might do, such as helping to provide ethical frameworks to guide our use of the technology.

With regard to the latter, the golem case is again interesting. Sometimes it is asked in AI ethics how we should treat AI, given that it exhibits humanlike abilities. Philosophers debate the moral standing of AI in a secular context. But it is also interesting to consider religious thinking about how to treat an artificial entity. In the Jewish religious tradition, for example, rabbis have debated whether a golem is a person. They have agreed that golems lack a divine spark or soul, yet they are animate. How should they be treated? The rabbis have concluded that while a golem should not be considered a full person, it is still important to treat it with respect since the way we treat it determines the development of our own characters.[40] This argument is in line with contemporary philosophical work on this topic, in particular applications of virtue ethics, which can also be linked to Immanuel Kant's argument for why we should refrain from shooting a dog.[41] Kant did not think that dogs themselves have moral standing, but he did think that shooting a dog would be bad for one's own moral character and hence the treatment of human others. What matters, according to these arguments, is our own moral character, regardless of the features of the entity in question. The argument is also politically relevant in that if we don't know the status of someone or something, we better be cautious and treat it/them with respect regardless of legal or other status. This does not only apply to AI but also to humans and nonhumans who might appeal to our ethical responsibility.

Today, many religious leaders have offered statements on AI ethics. For example, the Vatican has released guidelines on the ethics of AI.[42] At first sight, the guidelines seem rather industry friendly and therefore can easily be interpreted as supporting current power structures. The document is a meeting of Rome and Silicon Valley, two powerhouses in the techno-religious power constellation. But what is true for the Bible itself and indeed for all religious and secular texts is also applicable to this policy document: other interpretations are possible. For instance, the guidelines' principle that actions must be for "the common good of humanity and the environment" can be interpreted in a way that calls for radical political changes, perhaps even revolutionary action. More generally, the history of the Catholic Church contains both conservative and progressive visions. While Church leaders have often defended traditional social norms and resisted social reforms, there are also liberation theology, pacifism movements, labor movements, and individuals that have pushed for social justice and revolutionary change. For example, the late Pope Francis was widely recognized for his emphasis on social justice and his focus on the poor, marginalized, and vulnerable; this vision was based on interpretations of Catholic social teaching and liberation theology.[43] This side of religious thinking could also inspire people working in AI ethics. In case the spell cannot be broken, it might be a consolation to religion critics that there are many kinds of spells, and some can change, and have changed, the world.

Being connected to this variety of political-religious dangers and potentials, AI thus becomes an ambiguous, daunting political matter. Sometimes it is an explosive, Medean cocktail of social issues, for example when it shows biased beliefs and discriminatory practices. Often it supports the existing social and political order. Occasionally it is used as an opportunity to criticize that order. AI is not just a technical instrument but, with its connection to various religious meanings, it also taps into the big issues and tensions concerning power in society. If AI is connected to religion, apart from whatever else it is related to, then it is also political. It is about oppression but also, at least potentially, about liberation. It is about our deepest aspiration to become gods and no longer have to work, but also about the political and social problems that have followed from attempts to realize this fantasy, problems that continue to haunt our contemporary world.

3
Do Not Abandon Me: Of Creatures and Creators

A stormy sky in the cold and rainy summer of 1816 was darkened by the eruption of Mount Tambora in Indonesia, the most powerful and destructive eruption in recorded human history.[1] A young woman, Mary Godwin, lay restless in her bed. Earlier that night, Mary, her lover the poet Percy Shelley, and her stepsister Claire Clairmont had joined Lord Byron and his doctor John Polidori in Villa Diodati on Lake Geneva. Confined to the indoors due to the incessant rain, they read poetry and German ghost stories. Byron proposed that they each write a ghost story themselves. Mary, Percy, and John accepted the challenge. That night, Mary listened to long conversations between Byron and Percy. In the darkness lit only by candlelight, they discussed the principle of life and a popular science topic of the day: Could a corpse be reanimated, perhaps if given warmth with electricity? The idea was fascinating and horrifying at the same time.

For days Mary was asked if she had a story, but she could not come up with anything. Now she had again gone to bed late but could not sleep. Ghostly moonlight leaked into the room through the shutters. Then she had a vision. In her own words, "I saw the pale student of unhallowed arts kneeling beside the thing he had put together. I saw the hideous phantasm of a man stretched out, and then, on the working of some powerful engine, show signs of life, and stir with an uneasy, half vital motion." A dead man had been vitalized. Mary noted that the "Creator of the world" had been mocked. The human creator, however, was not only successful but also fearful about what he had done. He hoped the monster would become a corpse again. But that did not happen. He would wake up to find that "the horrid thing stands at his bedside, opening his curtains, and looking on him with yellow, watery, but speculative eyes."[2]

Mary opened her eyes in terror. But now she had a story. Percy, who would soon become her husband, encouraged her to develop the idea at greater length. The vision she had at eighteen years old became a well-known book that still influences our thinking about science and technology and their moral consequences: *Frankenstein; or The Modern Prometheus*. A young scientist, Victor Frankenstein, develops a technology to give life to dead matter. When he reanimates a human corpse, he is horrified and abandons the creature. The creature, looking monstrous, is miserable and lonely and eventually takes revenge. Are modern technologies such as nuclear technology and AI not also a kind of monster we created but then feared? Have we gone too far? Have we committed hubris, as the ancient Greek stories of Icarus and Prometheus—explicitly referred to in *Frankenstein*'s subtitle—already suggest? Have we mocked God? And have we made technology "lonely," that is, have we abandoned our creations and failed to take responsibility for them? But how alien and monstrous is technology such as AI really, given that we created it? There is difference but there is also similarity. How should we deal with this distorted mirror image? And who is the real monster, the abandoned one or the one who abandons?

As these questions suggest, *Frankenstein* is not just a famous ghost story or a superficial warning about the dangers created by science and technology. It also raises deeper and more complex issues concerning the relation between creator and creature. And these issues can also be found in Jewish and Christian thinking and can be elaborated in a way that is relevant to thinking about contemporary technology such as AI. Can creators control their creatures? And are and should creatures be created in the image of their creator? What, exactly, is the relation between creator and creature? And what is the difference with, say, the relation between parent and child, in particular between father and child?[3] Should creators be responsible for their creations, as Frankenstein seems to suggest? What if a creator (father? mother?) abandons the technology they have created, running away from its unintended consequences and their responsibility for it?

Frankenstein is by far not the first narrative that addresses these problems. Genesis, for one, probably written around 2400 years ago, is about a divine creator—God—who creates humans in his image, but the creatures mess up and then there is the Fall.[4] Following their disobedience, eating from the tree of knowledge, Adam and Eve are expelled from the Garden of Eden and then abandoned by their creator, later called Father. The spiritual relationship with God is now damaged, if not broken. Humans become mortal and have to live in sorrow because in their disobedience they alienated themselves from their creator and then in turn were abandoned by their creator as a punishment. According to Augustine, one of the foundational figures in Christian thought, all human beings inherit this "original sin." The rebellious child is made to feel guilty and is punished by the father. The same happened to Satan. According to the Book of Revelation, Satan and his angels rebel in heaven, after which God sends them to earth. Satan is called "the dragon" and a "serpent."[5] Evil, perhaps, and a monster, but in any case rejected by the father figure.

Stories of abandonment resonate throughout the Bible, including the abandonment of innocents. The story of Job is a case in point. God creates a situation of abandonment for Job, where his friends leave him and accuse him of abandoning God. Job, who claims he is innocent, certainly feels as though God has abandoned him when he needed him most. But why would God abandon him? He cries out to him but God does not answer.[6] Psalm 22 reads, "My God, my God, why hast thou forsaken me?"—words that later are repeated by Jesus on the cross.[7] Even Jesus, seen by Christians as the son of God, feels abandoned by his Father. The relationship between creator and creature remains metaphorically linked to the father-child relationship, where the child does something wrong (or perhaps refrains from doing something), is abandoned and punished by the father, and in the end is made to feel guilty for the rejection. (Mothers are not in sight here.)

Thus, there were already stories of disobedience and rebellion by creatures and monsters—and the resulting abandonment by their creators—long before we started to write tales of robots and AIs that run wild. And in all these cases, the creator and "father" is at least partly morally responsible for what happens. He sets up situations for his creatures and then punishes them. Yet there is also the free will of the creature. Humans have the freedom to obey or disobey. Augustine stressed that humans were given free will, including the freedom to turn away from God. Without freedom, disobedience and rebellion is not possible. Similarly (but not the same), without *automation* technologies, which can do things by themselves, we would not have the control problems we currently have with AI. It is their automation qualities, their agency, and hence their "freedom" that give rise to the control problem on the part of the creator. And if creators did not abandon their creatures, that is, if developers of technology did not try to avoid responsibility for what their creations do in the world, there would be far fewer problems. In terms of the ethics of creating AI, robots, and other automation technology, then, the moral of the story is not only that our artificial creatures should be obedient to their creators (i.e., humans), for instance in the form of following rules, such as Asimov's famous Laws of Robotics, which demand that robots obey orders given by humans. There is also a lesson for the creator, which is to think about how much automation you want and create your technology in a responsible way that takes care of the consequences. Do not abandon your creation. Otherwise *you* are the monster.

This message also emanates from the stories about that other famous artificially created human and monster that I already mentioned earlier, the golem. The golem is a creature that figures in Jewish folk narratives, and its name in Hebrew means "unformed," "shapeless mass." Like Adam, who according to the Talmud (Tractate Sanhedrin) was also initially created as a golem before getting a soul, the golem is created from clay or mud. But unlike Adam, unlike contemporary AI, and very much like Frankenstein's monster, the golem cannot speak. The monster is mute, but animate(d). In the Middle Ages, it was believed that golems could be brought to life by magic. In the nineteenth century, Jewish writers wrote about the golem of Prague. It is said that Rabbi Judah Loew ben Bezalel created a golem out of clay from the banks of the Vltava River (in what is now the

Czech Republic) and then brought it to life by means of Hebrew incantations. Supposedly, the golem was meant to defend the ghetto from antisemitic attacks. But there are more stories. Sometimes it is said that the golem fell in love and became a monster and murderer when rejected. In these stories, however, the rabbi manages to turn the golem lifeless again. In contrast to Frankenstein's monster, the golem is back under human control. It is said that its body is stored in the attic of the Old-New Synagogue of Prague, Europe's oldest standing synagogue.[8]

To link the story of the golem to AI is not much of a stretch. AI is a kind of golem, an animated artificial entity, but one that can speak. There are even similarities at the "technical" level. The specific combination of Hebrew letters needed for animating the golem can be seen as a code. AI, like the golem, is capable of "animating" machines by means of code and coding. And again we encounter the creator-creature problematic. The creature (i.e., AI) is seen as dangerous and rejected by its creator. But can AI be controlled, like the golem? Can coding tame the unintended consequences of AI? Can it be switched off, if needed? Or will AI never become mute again?

Is this a far-fetched comparison? Someone who considered the golem myth highly relevant to thinking about contemporary technology was none other than MIT mathematics professor and founder of cybernetics Norbert Wiener, who used it in the title of his 1964 book *God & Golem, Inc.: A Comment on Certain Points Where*

Cybernetics Impinges on Religion. Wiener was a scientist but he also had an interest in religion and claimed to be a descendant of the medieval Jewish philosopher Maimonides (Rabbi Moses ben Maimon). After presenting his ideas on machine learning, Wiener asked questions about what he called the "game between the Creator and a creature."[9] If a creator is in control and omnipotent, can a creator play games with his own creature? Maybe if there is some distance, if the creature can do its own thing—even rebel. And think again about the problem of freedom and control: some freedom on the part of the creature seems necessary. But does this mean that God is not all-powerful? And if God makes humans in his own image, what does that mean for machines that replicate themselves? If God makes humans in his image and humans create machines in their image, Wiener asked, then what is the image of a machine?[10] Would this be a copy of the original or would it depart from that?

Wiener also stressed human responsibility for machines. He mentioned the golem but also Johann Wolfgang von Goethe's story of the sorcerer's apprentice to discuss how cybernetics gives us new powers and new rituals, but we don't know how to handle them and things get out of hand. Like in *Frankenstein*, the creature is out of control. This is our fault. For Wiener the problem was that today we use the magic of automation for personal profit or to "let loose apocalyptic terrors."[11] But in contrast to the sorcerer's apprentice, who is saved by the master when he returns and tames the technology, this time nobody will save us. Wiener did not say why, but we can venture that there are two possibilities: either God won't save us or there is no god that could potentially save us.[12] Either way, we should not be waiting for a deus ex machina to save us (or for the master sorcerer to return home) but instead take responsibility for our own creations. We must develop technology responsibly. Such responsibility should not be outsourced to bureaucracy—Wiener mentioned Hannah Arendt's concept of the banality of evil here—or to the machines themselves.[13] We are responsible for our technologies and should assume that responsibility. Moreover, not everything should necessarily be delegated to machines, even if it is possible to do so. Wiener argued that we should think of how to divide agency and labor between humans and machines. In analogy to Jesus's famous words about the division of power between the emperor and God, Wiener wrote, "Render unto man the things which are man's and unto the computer the things which are the computer's."[14]

Wiener's questions are helpful when it comes to reflecting on contemporary AI. In what sense and to what extent is generative AI, for instance, an "image" of the human? What is the same and what is different? And what is the status of AI's creations? How much control do we have and how much control *can* we have over what it does? How much "freedom" do we want to give to our machines? When does our gaming of the technology end and where does AI's gaming of *us* start? How can we disentangle AI from ruthless profit making and military uses that could unleash the "apocalyptic terrors" Wiener talked about? How can we ensure human responsibility and accountability for AI? What AI myths facilitate evading that responsibility? How can we avoid seeing AI as a being on its own, a destiny, or a deus ex machina that will save us? And what would be a good division of labor between humans and generative AI, for example, when it comes to writing?

The relation between creator and creature—and related questions about freedom and responsibility—have always been under investigation in Judeo-Christian thought. Let's return to the story of Job, also mentioned by Wiener, which features the devil.[15] God and the devil play a game with Job: Will Job still refuse to curse God if he is tormented by the devil? But, Wiener asked, how can God play a game with the devil given that he created the devil? And if God is omnipotent, why would Satan's rebellion have any chance? Why does God need to prove his point at all? There is something paradoxical about the rebellion of the creature against its creator since the creature is linked to the creator, who creates the creation in his image.[16] How, then, is a rebellion possible? The only solution to this paradox is to argue that the creator gives freedom to the creature and then no longer has full control and power over it. Thus the creator is no longer omnipotent. Humans use their free will. Satan plays his own games and can rebel because he is a fallen angel that has freedom. But this means that God is no longer all-powerful. Wiener seemed to endorse that conclusion when he wrote, "God is something less than absolutely omnipotent."[17] Moreover, now the image starts to differ from the original. Humans no longer look entirely like God. God is no longer an all-powerful designer who shapes the world entirely according to his image. The image is distorted. There is a gap. There is similarity but now also difference. The creature appears uncanny. It becomes a monster. It can be played with or abandoned. But it can also rebel.

While this loss of power on the part of the creator may be difficult to accept theologically, that is, when it comes to thinking about God, in the case of technology it is easy to see how, through unintended consequences, it can get out of control and no longer correspond to what its creators had in mind. Technology is designed by humans, and there is certainly some control on the part of those who create and employ it, but humans do not fully control what technology is and does after its creation. In contrast to what some apocalyptic narratives about AI suggest, AI may not be(come) a kind of Satanic rebel, as if it were a humanlike agent with humanlike freedom. But at the same time, it is not fully under its creators' control either. Once the technology is used widely, AI developers no longer have control over its consequences in the world. They neither are omnipotent nor do they fully control what the technology is going to be, mean, and do in society. There is a gap between the design and the *use* and impact of technology in the world, sometimes in unpredictable ways. AI has monstrous, unexpected, and uncontrollable sides.

This seems especially the case with contemporary machine learning AI, where output and decisions cannot be reduced to the rules given to it by humans. Consider the unpredictable and seemingly uncontrollable output of large language models. Let us return briefly to the golem story. Perhaps the biggest scandal, so to speak, of ChatGPT and similar programs is that they are not mute. Now, this was intended. They are supposed to "speak," of course, they are supposed to use language and produce text. And LLMs animate chatbots, for instance. They are meant to simulate conversation; this is what the AI magicians aimed at. And the most scandalous, fascinating, and at the same time scary thing is that they work. The golem has become alive. Out of the silicon "clay" and data, creatures have risen. There is animation. The golem now speaks. But this scares us, and it scares its creators. It mocks not only God but also us. It works better than expected. We cannot control or predict the outcome. And what if the creature stops being polite and rejects us? Speaking is political. The one who is able to speak is also able to speak back. There is the possibility of contest and rebellion. There are many other unintended behaviors and unintended consequences. And there is some abandonment and irresponsibility on the part of the creators. Big Tech might run away from the consequences of its creations, take its hands off things, and let the creatures make their mess. They might fire their ethics department, for example.[18] Perhaps they outsource AI ethics to professionals, bureaucrats, and politicians, who are then supposed to deal with the abandoned monster. Or they might ask us to wait for a deus ex machina, who, or which, will solve everything. They might put their hopes in AI itself.

To avoid this banality of evil and abandonment of the creature, AI's creators need to take responsibility for their creations. In line with Wiener, we can argue for the responsible development of AI so that "rebellion" of the technology is avoided through new rules, involvement of stakeholders, and more ethical design controlling what our creatures do. If we regulate and code the creatures in the right way, if we *give them ethics*, they will not rebel, they will be AI for good—or so we expect and hope.

Yet even if we create a creature in our image—if we create "human-centered" AI—and try to make it more ethical in the sense of trying to give it a *human* ethics using our human image or at least the better part of it, there are limits to what we can do as human creators to ensure that technology is and remains good. There are limits to control on the part of the creator. In the human case, this is so because God has given humans free will. In the context of automation technology, there is no free will and no (bad or good) intention, but there are limits to control since there is automation and there are unintended effects. In the case of AI, there is more automation, learning, and development than ever. Once you let technology do its own thing, it might do things that are not right and not good. "Rebellion" in this sense is always possible and even likely. This implies that the creator of technology is in principle not omnipotent. By creating automation technology and giving it autonomy, we have also created potential rebellion and competition and the possibility that humans do not win the game. This is *Frankenstein* and the golem again, and a familiar science-fiction theme: consider movies such as *Blade Runner* (1982), *The Terminator* (1984), *I, Robot* (2004), and *Ex Machina* (2014). But it is also a reality. There may not be an AI Frankenstein monster or AI golem, at least not literally, although AI-powered robots increasingly look like them. But technology partly does its own thing, and humans are not omnipotent as creators. There will be bad consequences, and we cannot avoid all of them at the point of creation. We need measures to deal with these consequences in case they cannot be avoided; we need regulation. And—a point that receives less attention in discussions about AI ethics and regulation—we need users to act responsibly. At the same time, creators of technology have the opportunity to create technology as much as possible in their own image (which arguably should reflect the best of humanity and its diversity, not the image of a select club of tech people and owners and their biases and interests) and take responsibility for it. To abandon our creatures is irresponsible. But in light of the creator-creature problem, we should acknowledge that there is never full control and full mirroring between the image of the creator and that of the creature. The creator-creature tension and gap remain.

The only way to avoid these conclusions, that is, to avoid creator responsibility (while recognizing the limits to creator control), is to see technology not as a creation but as something else, for example as something that evolves on its own or that has absolute alterity: a total other, something that has nothing to do with us. It is sometimes said that AI is alien. This avoids creator-creature problem(s) since there is no longer a quasi-parental relation between creator and creature; issues such as abandonment and rebellion do not apply. There is only difference and distance; there was never a bond in the first place. Yet this approach also alienates technology from humans, falsely suggesting that there is nothing human in technology or that technology is not at all made in our image. This picture is wrong; after all, we *are* its creators. AI is *not* alien in this sense. If there is evolution, it is an evolution in which we, the creators, have a hand. Perhaps there is coevolution and we evolve influenced by our creations, which is also something that happens in the parental relation. There is a kind of feedback and learning, a mutual shaping. But we are still the creator-parents. And even if our image gets distorted when it takes the form of AI, it is still present in the technology since we designed it. Therefore, that technology can never be completely alien to us as we remain its creators. And this means that we incur the related responsibilities and problems. As long as we stay within creational thinking, which certainly makes sense when it comes to technology, its paradoxes and challenges are bound to remain with us. Therefore, unless we manage to find an altogether different relation to technology, lessons from the creational religious traditions that reflect on these creator-creature problems remain relevant (and, arguably, have set us up in this way of thinking and with these problems in the first place). Whatever we may think of religion in general—or indeed about a particular religion such as Christianity—we can use the religious traditions and their theologies to further investigate our relation to technology.

The problem concerning omnipotence, for instance, can also be framed as the problem of evil. Here another earthquake sets the scene, about sixty years before Mary Shelley's vision. From nineteenth-century Romanticism we turn to the earlier age of Enlightenment. On the morning of All Saints Day, November 1, 1755, Lisbon was destroyed by an earthquake and a tsunami. An estimated 60,000 people died in Lisbon alone, many when crowded churches collapsed on them. Philosophers all over Europe discussed how this could happen. Wasn't God supposed to be good? Voltaire asked how God could allow such evil, given that he is all-powerful. He argued against the view of another philosopher, who died before the earthquake and who played a role in the history of AI, Gottfried Wilhelm Leibniz. In his *Theodicy*, Leibniz argued that God created the best possible world. That world has evil in it, but after the creation, God no longer intervenes. We should have faith and hope for the best; there is an overall universal good, rooted in the good divine creator.[19] Voltaire, by contrast, mocked Leibniz's view that we live in the best possible world and argued that we should alleviate suffering and eliminate evil. Instead of trusting the divine gardener, Voltaire suggested in *Candide*, we should attend to our own garden. We should take responsibility.[20]

Some developers of AI and their companies seem to embrace Leibniz's view of creation: they seem to think that it is not the developers' task to deal with the consequences of their creations. The developers have to aim for the best possible technology. Afterward, they let it go and are not supposed to intervene. There may be problems, but a world with AI as it currently exists is still the best possible world. To deal with the bad consequences is a task for others—for users, consultants, regulators, and so on. They are optimistic about the end result of technology and ask us to trust them as creator-developers. Following Voltaire and Wiener, however, we should respond that developers should also take responsibility for their creations and eliminate potential bad consequences both at the point of creation (Wiener) and later, when the technology turns out to have all kinds of consequences in society (Voltaire). Creators of AI should not be indifferent parents, distant feudal lords, or gods who don't care about their creations, and if they are such careless parents, lords, or gods, then we have no reason to trust them. They might even be maleficent. We should limit their ownership or even take it away. In addition, as users and as a society we should also exercise responsibility at the point of use and when bad consequences happen. We had better make sure that—whatever a creator does or does not do—we take care of our own garden and cultivate AI for good. Users and civil society should play an important role in this.

Leibniz, however, was not only a philosopher but also a mathematician. Back in the seventeenth century, in 1666, the twenty-year-old polymath published a dissertation, *On the Combinatorial Art,* in which he outlined a theory for automating knowledge production via rule-based combinations of symbols. He dreamed of a machine that could calculate ideas, a thought that was inspired by the Catalan mystic Ramon Llull, who in turn was influenced by Jewish Kabbalism. The machine was meant to solve disputes between philosophers. "When there are disputes among persons," Leibniz wrote in *The Art of Discovery,* "we can simply say, Let us calculate, without further ado, in order to see who is right."[21] Later he constructed a mechanical calculator. But on this topic, too, Leibniz and Llull had a critic: Jonathan Swift. In his 1726 *Gulliver's Travels,* he parodied Llull and mocked the idea of a machine that could generate language and write books, and even the idea that machines would ever be able to understand the meaning of words. Swift described "The Engine," a fictional machine that generates permutations of word sets. Today, when we discuss generative AI, we have this debate once again. On the one hand, a lot of progress has been made by means of LLMs. Now widely used (even by academic philosophers), they seem to understand what we ask them and what they say, and sometimes they may even appear conscious and sentient. But this is a kind of magic illusion. These technologies do not understand meaning and are not aware of what they do; rather, they merely make predictions on the basis of statistical analysis of data. We have a new version of The Engine, a machine that uses computation to create wonder, illusion, and deception, a machine we can even talk to, but still a machine that does not really understand us.

In the next chapter, I will say more about magic and machines. In the following one, I will ask whether chatbots are the new priests or gods.

4
Make Me Wonder: The Magic of Automata and Aladdin's Lamp

It's autumn 1769. Maria Theresa, the empress of the Austrian-Hungarian empire, who is very interested in the sciences, invites Wolfgang von Kempelen to court to watch a performance of "magnetic games" by the French scientist Pierre Joseph Pelletier. Pelletier's performance mixes science with magic tricks. Kempelen, a Hungarian civil servant who had been appointed counsellor to the imperial court, watches attentively. He also does scientific investigations. He has his own workshop and is a passionate inventor in his free time. With his reputation rising, he has been commissioned to design the waterworks for what is now Bratislava castle. Now he is asked by Maria Theresa what he thinks of the performance. To everyone's astonishment, Kempelen claims that he can construct a more surprising machine. The empress challenges him to do so.

In 1770 Kempelen is ready, and the chess-playing automaton, nicknamed the Turk, debuts at Schönbrunn Palace. It consists of a wooden cabinet on wheels with a figure of a man with a turban (hence the nickname) who stares down at a chessboard and holds a Turkish pipe. Kempelen announces that he has built an automaton chess player. Opening one of the cabinet doors, he shows a mechanism of wheels, cogs, and levers. Then he asks a volunteer to be the opponent and the Turk starts to play, moving its head and then moving one of its chessmen forward. The Turk defeats the volunteer. But not only that, it also performs the so-called knight's tour, a chess puzzle that requires the player to move a knight to occupy every square of a chessboard exactly once. The Turk solves it. The audience, including the empress, is astounded. It seems magic.[1] The Turk will tour Europe and be demonstrated for Benjamin Franklin, Frederick the Great, and, after Kempelen's death, Napoleon Bonaparte.

Kempelen's machine was a contraption controlled by a human operator hiding in it—the part that Kempelen *didn't* show to the audience. It would take until contemporary times for a supercomputer to defeat a world chess champion. Nevertheless, both the mechanism and the trick were impressive, and the mechanical Turk became one of the most famous cases in the history of automata, which has always been and still is a history of science and magic, ingenuity and deception.

That history of automata is fascinating in itself. It stretches from Greek automata such as the mentioned wine-pouring maidservant to contemporary robotics and AI. Each time, inventors try to impress and deceive their audience, including donors and investors. Each time, the pretense is to cross the border between inanimate and animate, between dead and alive. Each time, there is a mixture of science and magic. There is a performance. Or to put it in religious terms, there is a ritual. Big Tech also has its rituals. When in May 2024 OpenAI CEO Sam Alman announced the new model ChatGPT-4o, he said that the company would present something that "feels like magic."[2] As usual, teasers were followed by a magical reveal. The world still watches in awe when contemporary Pelletiers and Kempelens show off their magic technologies—AI systems and robots—that can see, speak, and hear: automata that increasingly seem alive.

There is a good anthropological explanation for our fascination with these border crossings by way of artificial entities. Building on Lévi-Strauss's work, Minsoo Kang explains that communities construct a reality made up of binary categories, but in rituals there is a temporary release from this strict order, which in turn is meant to reaffirm and maintain the order.[3] In their rituals, magicians make it seem as if an object is coming alive. The automaton, inanimate and animate at the same time, crosses the borders of those categories, inhabiting a spooky zone in between. The automaton is undead. It is "a categorical anomaly":

> It is an artificial object that acts as if it is alive. . . . It appears to leave the hands of its mortal maker and take on a life of its own, animating itself to mock the idea that the power of creation belongs to God alone. The binary categories of living/dead, animate/inanimate,

> creature/object all break down in its wake, as it moves from one to the other, mesmerizing and terrorizing its beholders by turns. And its maker is not only an artist but a magician as well, perhaps an ambiguous creature himself who traverses the worlds to tap into uncanny power.[4]

The term "uncanny" is borrowed from Freud and lives on in contemporary robotics. He explored the term in his 1919 essay "The Uncanny," in which he describes the encounter with things that are at the same time familiar and eerie.[5] Some objects such as dolls and automata are situated on the boundary of the familiar and the strange. Such objects lead to discomfort or dread. In 1970 the Japanese robotics professor Masahiro Mori took up the concept again when he described what he called the "uncanny valley" phenomenon.[6] As the appearance of a robot (or a doll for that matter) becomes more human, observers respond emotionally positively to the robot, even empathically. But once it becomes *almost* human, it begins to look strange to people. Looking *too* human, it produces an uncanny feeling. We then respond in a similar way as we do toward corpses or zombies, which look very human but also create revulsion. According to Mori such robots make us aware of our mortality and trigger fear of reanimation of the dead. The topic continues to spark interest in the robotics and robot ethics community today.[7]

What is going on when we interact with automata is also often attributed to animism, which is also mentioned by Freud. We tend to attribute a soul or spirit to the objects around us, especially when they are interactive—all in spite of modern science dismissing their existence. This is one way in which we, to use Bruno Latour's words, have never been modern.[8] We may think we are modern, but in terms of perception and emotional experience, we are still to some extent inhabiting an enchanted garden. We are taught to think like scientists but we experience the world like Alice in Wonderland. It is an experience and way of seeing the world that has been common to all nature religions and therefore present for most of the time humans have been around. Monotheism has tried to erase, and sometimes partly incorporate, this religious experience, but has only been partly successful.[9] When we are confronted with automata and other wonderous technology such as AI, we find ourselves again in a nonmodern Wonderland. We fear and are at the same fascinated by the new magic garden, in which previously dead objects now speak. There is a plurality of spirits.[10] Animism is alive and well. This is also the case for AI. As Stef Aupers has argued about technoanimism, in AI subjective characteristics are attributed to lifeless objects, and this animism is not just something that happens unintentionally or only after development; rather it is "the logical result of a conscious and purposive implementation of human characteristics in these objects."[11] In other words, AI animism is designed from the start. Automating objects and giving them speech, AI designers fulfill the dream of a truly ubiquitous technology that turns the dead, artificial environment into a jungle full of life, not unlike natural environments as understood by pagans and shamans. AI can thus be seen as partly the result of a postmodern, romantic re-enchantment project. As Aupers has rightly suggested, rationalization, seen by Weber and many technology critics as characteristic of modernity and modern technology, can also turn into *re*-enchantment rather than (only) disenchantment. AI, while often interpreted in rational terms, is related to *both* sides of modernity's coin.

Another link to religion that is mentioned in Kang's description of the automaton as a category anomaly is again the creational one, where the inventor-magician "plays God" when trying to imitate the process of creation or, rather, when trying to *become* a creator. Here the problems of the creature in the creator-creature relationship are "solved" by the creature (i.e., humans) becoming creator, which, as we have seen, then leads to new creator-creature problems between humans and their creations. Whether or not this can be interpreted as a Freudian killing of the father (i.e., the infamous Oedipus complex), it is certainly a new constellation. The creature is now, at the same time, creature and creator. From an ancient Greek point of view, it is a case of *hubris*. It disturbs the divine order. Humans are not supposed to become (like) gods. From a Christian point of view, it depends. Some argue that humans should not interfere with God's creation and that life is sacred and in the hands of God. They see creation as a rebellion. Others argue that creating technology is part of human stewardship over the earth and that it is morally needed, for instance to relieve suffering and enhance quality of life. Some even support human enhancement from a Christian point of view. For example, theologian Ronald Cole-Turner has argued that God has given technology to humans to improve the human condition and fulfill the potential of God's creation.[12] The doctrine of *imago Dei* is thus interpreted as implying that we should become creators ourselves, or at least cocreators, as another theologian I mentioned earlier, Hefner, has argued.[13] We create machines in our image: *imago hominis*.[14]

In the meantime, the magic shows go on. Each time there is a combination of science and magic, technology and trickery, innovation and religion, artifact and legend. And each time, the creation and experience of these technologies taps into a multitude of meanings that can easily be connected to the rich, varied, and colorful history of religious ideas and practices.

Already in medieval times automata were developed in a hybrid context of magic, alchemy, and Christianity. For example, in England the thirteenth-century Franciscan friar and philosopher Roger Bacon, seen as one of the pioneers of the scientific method but also at the same time as a wizard, developed a fascination with astronomical clocks and allegedly made a brazen head that could speak. Bacon was a disciple of Bishop Robert Grosseteste, a philosopher and theologian who lectured in mathematical physics at Oxford University and who is said to have made a brazen head that could predict the future. In the same century in Germany, the Dominican friar, philosopher, scientist, bishop, and saint Albertus Magnus (Albert the Great) is said to have created a head that could talk. According to Alfonso de Madrigal, also known as El Tostado, Magnus worked for thirty years on it, and the metal man would answer all his questions until his student Thomas Aquinas destroyed it because he was annoyed with its murmuring. Much later, in the eighteenth century, Kempelen also developed an interest in building a speaking machine.

Magnus's head is often seen as merely legendary. It might have never existed or perhaps it was a device that tricked people into believing it spoke. But later, concrete steps would be taken toward a science that lay the basis for contemporary artificial intelligence. Leibniz would envision a universal symbolic language—a symbolic system that would represent all human knowledge—that he thought would enable humans to create an automated system capable of reasoning and problem-solving. The concept was a precursor of ideas and techniques that would form the basis of programming languages and AI systems. Today, we finally have talking machines: a new kind of magic enabled by natural language processing.

*

Current robotics and artificial intelligence stand in the tradition of magic automata in at least the following ways.

First, these technologies are often aimed at imitating humans in a way that impresses and deceives audiences. Humanlike robots such as Ishiguro's androids or Hanson Robotics' robot Sophia are presented in public performances and demonstrations much like Kempelen's. The roboticists and their teams are not only scientists but also artists and magicians. And when OpenAI demonstrates its new chatbots, this is not unlike the unveiling of a new speaking head. Investors, consumers, and policymakers have to be attracted and entertained. Created by well-paid techno-wizards, the new devices are surprisingly humanlike in their appearance (e.g., Ishiguro's robots) or speech (e.g., OpenAI's GPT series). Contemporary philosophers may well be annoyed by the endless textual murmurings of technologies such as ChatGPT, and some scientists may take up the role of skeptic and spoilsport revealing the workings of the machines and pointing to their limitations, but the technologies are popular and widely used.

Second, these machines do not only fascinate us but also trigger fears as they cross boundaries between object and subject, dead and alive. They are category anomalies. Ishiguro's robots are said to be uncanny, and chatbots have been called conscious and sentient. Consider the story of Blake Lemoine, who was a software engineer at Google. He had been working on the development of the large language model LaMDA for months. After many dialogues with the technology, Lemoine claimed in an internal document that the chatbot was sentient. After his claims were dismissed by the company, he made them public. Moreover, he wanted the personhood of LaMDA to be recognized. This sparked a public debate about the moral standing of LLMs. The technology thus creates confusion. And this is exactly part of the aim of the engineer as magician: to make it seem as if the chatbot is alive, conscious, sentient, and so on. The moral confusion is a sign of the success of the magician. (In this context it may be interesting to know that Lemoine identifies as a priest and follower of Discordian religion, which worships the goddess of chaos and is known for its interest in trickery.)[15] Today the confusion continues when some scientists call for testing systems for consciousness and propose "AI welfare" policies.[16]

Third, the project of creating humanlike AI and robots can be seen as playing God, or rather *becoming* creator, now creating artifacts in *our* image. We are the sorcerer's apprentices that experiment with creation. All the potential risks and negative consequences have been attributed to this hubris. Or is it our duty to continue creation? In *Humanity's Last Stand,* Nicanor Perlas argues that AI can be viewed as a continuation of a broader spiritual and creative process and can be a force for good.[17] Computer scientist Ray Kurzweil, using explicitly religious language, has argued that the development of AI is a continuation of the creative process that began with the evolution of life on earth and is part of a spiritual evolution of intelligence. In *The Age of Spiritual Machines,* he predicted that the problem of mortality will be solved since humans will merge with intelligent machines and that intelligence will move beyond Earth and change the universe.[18] In *The Singularity Is Near* he argued that technology can help us to transcend the constraints of the physical body, reach expanded consciousness, and merge with machines.[19] After the Singularity, the universe will wake up and be filled with divine-like superintelligent entities. Developers of AI can thus understand themselves as contributors to a larger cosmic evolution of intelligence.

Fourth, in Silicon Valley technologies, not only is science present but all kinds of other ideas come together, some of which connect to magic, alchemy, and religion. In *TechGnosis,* Erik Davis argues that there are connections between digital technologies and occult practices: understood as technocultural hybrids, they are not mere things but also link to the religious imagination.[20] Like Fred Turner, he claims that the mystical and spiritual counterculture of the 1960s influenced the development of digital technology and promoted a magical relationship to it.[21] As he puts it in the afterword, we have fantasies about a world in which desire is instantly satisfied and in which we are served (see chapter 2), and there is widespread technological enchantment as we come to understand algorithmic agents as having minds of their own. But as usual our fascination is paired with fear. Our fascination with digital technologies may be animistic, but it is "a kind of anxious animism" since we feel we are surrounded by many nonhuman agents and invisible forces beyond our control.[22] We are wondering but we are also afraid. We want the enchanted garden but when it actually seems to happen, we stagger and panic.

But perhaps it helps to remind ourselves of that other meaning of magic mentioned in my first point, that is, "Technology is a trickster," as Davis puts it.[23] Our relation to it might be shaped by magical thinking in a spiritual sense, but it's also magic in the sense of the art of deception. What Bruce Tognazzini has said about the design of human-computer interfaces is still the case today for our machines: the principles of magic and showmanship apply.[24] That being said, sometimes the magic is indeed experienced as a lot less fun and rather scary. What was first experimental hippie technology later became something a little darker, perhaps even demonic. We fear the unpredictable consequences of the AI revolution. And computing and data are now about money, big money. As users, we feel we are manipulated and used. And we don't know what will happen next. We feel that developers have unleashed an invisible, powerful, and unpredictable new force into the world: AI. We do not only fear; we also don't know what we fear. We suffer from AI anxiety.

Finally, now as in the times of Magnus and Kempelen, one of the most technically challenging and culturally fascinating projects in the history of magic automata is to create machines that can speak. And today, we finally have it. Behold our new talking heads in the form of LLMs—whether in the form of disembodied chatbots or as software that makes robots and other artifacts speak. And this time, it is very hard indeed to dispel their magic. It becomes increasingly believable. This is scary but at the same time we do not want to break the spell; we love wonder and mystery. We invite and welcome the techno-tricksters. And we increasingly depend on them and their fascinating but also uncanny technologies.

In what sense, then, can it be said that today's AI, especially in the form of chatbots and text generators based on LLMs, are *magic*?

First, as mentioned LLMs are part of a game of creating illusions and deceptions. They do not possess consciousness, sentience, mind, intention, voice, and so on, but it *seems* like they do. The magician-designers make sure that we are under that impression. Their aim is not only entertainment but profit. If people get drawn into these technologies, they can sell premium versions of the technology and then sell the resulting conversation data to advertisers. The new automata can also attract new investment. The best developers are hired to perform their magic tricks and perfectionate the illusion by technical means and means of media theater, which today increasingly and perhaps predominantly takes place on the stages of social media. "It's pure magic!"; "Our LLM is sentient!"; "We will soon achieve consciousness!" The words of the magician are part of the show and contribute to its success. And the words spread fast and widely. The magic show's reach is global and knows no boundary. The tech market is global, too.

Second, AI can be used to create wonderful worlds. Image generation by means of AI, for instance, can take us to all kinds of wonderlands. Today we are all Alice. AI image generators and AI-generated virtual worlds are the white rabbits that invite us to follow them down the rabbit hole. As guests of techno-shamans, we roam worlds we could never have imagined before. In the end, the magic of AI is not restricted to virtual worlds but also transforms the physical world into a fascinating and sometimes scary wonderland with all kinds of anthropomorphic creatures such as robots and talking devices. A world where there is no longer a sharp boundary between inanimate and animate, dead and alive, human and nonhuman. Previously dead objects become undead zombie objects. And what is most frightening of all: previously living beings become machines themselves, used by the vampires of techno-capitalism.[25] But who stops the sorcerer-apprentice and cleans up the mess? Who can teach the sorcerer-apprentice? Who can enforce new boundaries?

Third, technologies such as AI-based chatbots do their work in a way that we, as users, do not fully understand. For most users they are black boxes. There is mystery. Even the magicians-technicians do not have fully transparency. LLMs are built on deep neural nets with millions or even billions of parameters and many layers;

this makes it very difficult to interpret how a particular output was generated. We cannot simply "read" their "internal states" and explain the output. LLMs also generate unexpected outputs, thus displaying what can be seen as emergent behavior. What LLMs come up with is unpredictable. Furthermore, often the technology is hidden from view. We use an app but do not always know that it is powered by AI. And what we cannot understand attracts and worries us. There is agency but we don't see or know it. We are surrounded by technoghosts. The technology brings back the mystery in a world that we thought was purged of magic but was actually never disenchanted in the first place. In spite of our modern thinking, our experience always retains magical elements. Now AI adds to the mystery and wonder—and terror. We do not understand our own minds and we do not even understand our own creations. Our romantic minds revel in this.[26] They bask in the glow of their fascination with, and fear of, AI's magic.

Fourth, AI-based chatbots are much like Aladdin's magic lamp. Promised to be the ultimately general-purpose technology, they are designed in such a way that we feel the AI genie fulfills all our wishes and desires. This is most clear when it comes to our thirst for knowledge and companionship. We just have to ask. We just have to rub a bit, type a quick prompt, and behold, there, magically the answer appears, written by an invisible AI genie. The chatbot genie releases us from ignorance and loneliness. We can know anything now. We can talk. We can write. And maybe the AI lamp will make us rich and powerful, like Aladdin's did. AI promises to change the economy, to further growth and wealth. We just have to use AI and our neoliberal wishes will be fulfilled—at least if we somehow survive the games of trickery and betrayal in the business. The AI economy becomes an exotic land in which everything seems possible but where competition is stiff and no one can be trusted.

AI can also be seen as a kind of alchemy, understood in a broad sense. In ancient times, alchemy aimed at the transmutation of base metals into gold. Today, AI is predicted to achieve the transmutation of raw data, in themselves pretty base and boring, into shining and precious outputs that can be turned into profits. Large amounts of data are turned into something of value and sometimes something beautiful. AI gives us new knowledge, or so it seems. AI might also help us in our spiritual journey, as it seems to support efforts at self-improvement.[27] There may even be the hope that AI can give us the so-called philosopher's stone, that it might help us to discover the elixir of life, enabling immortality (see chapter 7).

Finally, by drawing on a very large amount of data, LLMs do what had always seemed impossible since it was seen as exclusively human: they speak and write. With this feat, they cross a boundary that had been seen as definitive of humankind since Aristotle's time. Suddenly we are no longer the only entities with logos. We are amazed but also fearful. The self-image of humankind is shattered, and now we must define ourselves in a different way. Even the image of *language* is shattered; apparently it can all be automated. Its mystery has been cracked, or so it feels. No theories needed. Not even minds. Statistical calculation suffices. A new step in the ongoing Copernican revolutions that dethrone the human and now even our sacred baby or mysterious mother, language. The postmoderns knew that the human soul was gone but hoped to find it still in the wilderness of language and the darkness of literature. This hope is now also in vain. It seems that AI can do it all. Yet we are as much disillusioned with humankind and the humanities as we are fascinated by the magic of this new technology. We stand in awe. Perhaps we will worship it. As LLMs perform the miracles of speech and writing and appear all-knowing after they do their alchemy—transforming base data into the gold of knowledge—we wonder if AI is a new divine power, a deus ex machina that was initially created by us but that will evolve and eventually save us. The new speaking heads might turn out to be the mouthpieces of a new god and overlord, or perhaps the technologies are themselves new gods that will take humanity to the next level.[28]

In the meantime, new priests and popes are ready to mediate our relation to the new God, tech priests and prophets who today are already accumulating unprecedented wealth, influence, and power. Those who claim to know and make AI's future are carefully listened to.

As we will see in chapter 6, even AI itself is consulted like an oracle. But perhaps the most mesmerizing, addictive, or even seductive feat of contemporary AI is that it is not just one-way communication. There is the promise of conversation and companionship. AI magically seems to offer us one of those other things humans need, existentially and spiritually: relationship.

5
Talk to Me: Prayers, Priests, and Chatbots

The robot SanTO has the appearance of a saint.[1] It stands in a neoclassical-style niche, and its palms are turned upward. The machine is meant as a prayer companion. People can touch the hands of the robot and ask questions, and the AI system answers. It proposes prayers and quotes biblical passages and other texts related to the Catholic faith. The answer is not always directly relevant to the question—large language models often make mistakes—but some people don't mind and seem to find the robot helpful, for example, for making them think about their own answer. According to Gabriele Trovato, the inventor of the robot, the role of the robot is not to replace the priest but to help people pray.[2] There are also online chatbots for prayer.[3] For example, a Christian AI app promises that you can now "text with Jesus," thus offering "a divine connection in your pocket."[4] AI thus plays the role of priests or saints as mediator of the relation to God, or takes over the role of God. People confess to it, request things, ask for insight and advice, ask for help to reach particular goals. It has even been suggested that AI may help with prayer understood as inner dialogue (introspection, thinking) since its words and texts may stimulate this.

Many of these uses of AI should not surprise us. Humans have always used tools and techniques in their religious practices. But AI, like all technologies and media, is never just a tool. It also has unintended consequences. The influence of AI on our lives and thinking as a whole should not be underestimated. Beyond the explicitly religious roles mentioned, AI increasingly gets an important place in the life of the soul; it becomes an intimate companion. People may not always explicitly worship AI, but they do give AI-powered apps a central place in their daily lives—not dissimilar to what religious people do, or are supposed to do, with God. Like all our technology use today, digital technologies, and in particular phones, have become what Epstein calls "a feverish ritual."[5] We are hooked to our screens. Opting out of the ritual is difficult. Whereas religions recommend prayer a few times a day, we look at our phones nearly all the time.[6] We even get up with them and go to sleep with them. As Gábor Ambrus puts it, "spending with it the most beatific hours of the day including the first and last waking moments (before going to pee in the morning and after doing so in the evening) certainly qualifies as a life of prayer."[7] AI plays a role in these daily tech rituals.

Furthermore, in the form of chatbots AI offers companionship and unconditional support to those who feel lonely, need therapy, or simply want conversation with an "AI friend." AI may thus be perceived as having a kind of "pastoral" role: it may be experienced as helping and taking care of people, assisting or replacing human pastors. And is prompting chatbots such as ChatGPT not similar to praying to an all-knowing and potentially all-powerful God, from whom we expect a response that helps us, without us understanding how the answer came about?

This interpretation would certainly fit some of the etymology of the term *praying*, which originates in the Latin *precari* (to ask, to request, to beg). The term was initially linked to pre-Christian offering rituals and sacrifices because when requesting something from divine power and authority, one offers something in order to receive something. However, like with chatbots, it is always unpredictable what the divine power will do. Will it grant our request? And if so, in what way? Later in religious history, prayer becomes on the one hand something more inner, a kind of accounting of the soul (in Hebrew, *cheshbon hanefesh*), related to introspection and self-improvement. On the other hand, it becomes *oratio* and is thus more about speech and conversation than offering. Modern definitions of prayer stress communication and dialogue with the divine. For example, in *The Varieties of Religious Experience,* William James defined prayer as "every kind of inward communion or conversation with the power recognized as divine," which includes petitional prayer (requesting something from the divine authority) but also left the definition open to other kinds of prayer and other acts of communication.[8]

This communicative definition of prayer also opens up a potential role for chatbots. As we have seen, chatbots powered by LLMs can now generate text and speech that seem increasingly more humanlike and are responsive to input from users, thus simulating conversation. This enables praying to either God or the chatbot itself. People request for and beg the chatbot to help them and have a conversation with it. And they get answers. Maybe the meaning of the answers is not always clear to them; but just as with divine signs and answers, that is not seen as a hindrance. People interpret. They make meaning. They figure out what it means to them.

Prayer is conversation and communication. But that conversation and communication can take all kinds of forms. People who pray may express gratitude, for instance, but they might also demand things or even express anger toward God, like Job did. People may ask, "Why did you do this to me, why did you hurt me?" Or they might not understand what God is doing. If prayer today is essentially speech and communication, then many speech acts are possible. But which speech acts, which communication, is appropriate? That depends, as always, on the situation, but also on the relationship. Prayers always assume some kind of relationship with a divine power. The nature of that relationship shapes the content and constraints of the conversations. It provides norms for the communication, gives a form or "grammar" for praying. But what kind of relationship is involved in prayer and what relationship *should* be involved? What is the appropriate relationship to the divine?

Philosophers and theologians have different views on this. In the monotheistic religions this issue is of course connected with centrally important discussions about the nature and image of God and the human relationship to God. Theologies say a lot about this. They are all about how to speak *of* God and how to speak *to* God (although the latter has received less attention, as far as I can see).

We can understand views of the relationship with the divine as being modeled on secular relationships in some way or other. When people ask something of God as an authority, for instance, the relationship has a hegemonic, usually feudal character. The relationship is that between a master and a servant, a divine lord and a mortal serf. God is the Lord that demands things from those who serve him and grants requests from them—or not. He may act as an arbitrary and malevolent ruler or a benevolent one. If God is such a feudal lord, he watches us, judges us, and may punish us if we misbehave, if we violate his laws. We are under his surveillance. He may protect us against the forces of evil. But he asks for offers from us, taxes. He keeps an account. People might also complain to or about such a God, in the same way that serfs or slaves complain to or about their lord or master. They may rebel against their lord or master. And they may not understand why their lord or master does something.[9]

Another and related model for the relationship with the divine is that of a business partner with whom one has a contract. I give you something—I make a sacrifice, for example—but then I want and expect something in return. Or from the point of view of the god, I make sure the world order continues, I protect you and your family, I grant some of your requests, I forgive you, and so on, but then I need something from you: a sacrifice, good behavior, and so forth. This way of thinking is also still present in the monotheist religions, although perhaps less explicitly. And for some God is more like a friend or partner with whom one can have not only a symmetrical but also a personal and deep relationship. God is then a companion that relieves our loneliness and to whom we can talk when we need to. We then expect that God listen to us as a good friend or partner would. When we pray, we expect an answer from God like we expect an answer from our friend or partner. We expect and give love, loyalty, and care. Sometimes we may be disappointed, but we try again. We confess. We repent. We forgive. We hope God forgives us when we have done something wrong. We hope he doesn't ghost us and doesn't abandon us. We are in a friendship or love relationship with the divine.

A thinker who exemplifies the latter view is philosopher Martin Buber, who studied not only philosophy but also Hasidic Judaism and regarded his thinking as rooted in Judaism. In *I and Thou,* he distinguished two kinds of relations, attitudes, and ultimately ways of being in the world: I-It and I-Thou.[10] The former treats people and the world as things to perceive and use. The object is separate from the subject; the world is separate from the self. The latter relation, by contrast, is noninstrumental and nonobjectifying. It is about mutuality and genuine dialogue. The other is not separate; there is a connection. It is about a living relationship. Buber advocated for having the latter kind of relationship with the divine when he argued that God cannot and should not be addressed as an object or an abstraction. God is a *Du* (German for "you"), a Thou. Prayer is then a dialogue with another you, albeit a divine one, that is, not a mortal you but an Eternal Thou. In this theology, God is not a "something" to believe in but a "someone" to live with. In modern times ruled by I-It relations, Buber argued, we should reestablish our relationship with God in a Thou kind of way. Prayer helps with this; in prayer, there is not distance but communication and communion. Prayers are always dialogical in character.[11] In prayer, we can have a dialogue with the divine you.

This view of prayer and of God has been criticized by another Jewish philosopher, Emmanuel Levinas, who argued that God should not be seen as a friend or partner. We can never have a symmetrical relation to God. According to him, God is an Absolute Other. In *Totality and Infinity,* Levinas argued that this Other, in its infinite alterity, is beyond comprehension and calls us to ethical responsibility.[12] Instead of seeing prayer as a dialogical and personal relationship, it becomes here an ethical openness to the infinite, transcendent Other. In this asymmetrical situation, direct dialogue is impossible. Silence and listening (to the Other), not speaking, is then appropriate. Prayer is about hearing a call to responsibility. Perhaps the divine may appear in the concrete face of the human other (e.g., in a war situation), but the way Levinas wrote about the Other also suggests that he saw God as something absolute, an absolute that is still a "something": the divine becomes again distant, abstract, and perhaps also hegemonic. To the extent that God is a person at all, it sounds like I have to respond to the call of the Lord and do my duty as a servant; I better listen. There is huge distance to the Lord.

On the one hand and at first sight, comparing conversations with chatbots to prayer does not seem to make sense when one adopts Buber's or Levinas's views of praying and the relationship to the divine. Machines, one could argue, can never be the Thou or alterity that are required in this kind of relationship, since they are not persons or others in the first place. One can therefore never have a Thou relationship with a machine and a machine cannot be an Other. At best, AI can be a quasi-divine dialogue partner in an authority or business relationship. It can tell you what to do and give you something in return for your money or your data. But it can never be a friend, partner, or Absolute Other. And in contrast to my responsibilities in a relationship with a human other or divine Other, I do not have any ethical obligations that arise from my interaction with a machine. The machine does not issue an ethical or religious call. Similarly, one can criticize robot or AI priests for lacking the required properties for mediating the divine. For example, in an interview about the SanTO robot, a Catholic priest argued that robots lack a soul and are not persons, and that therefore they cannot take over the tasks of a priest.[13]

On the other hand, referring to the properties of machines and comparing those to the properties of God could be criticized in a Buberian fashion by saying that the divine is not about properties, is not about an "It" but about a relationship. If we take Buber seriously, one could argue, we need to focus on the quality of the relationship to the divine, not on properties and things. If the relationship to the divine is a good one, one that is like friendship or love, then at least in principle AI *could* mediate and support this relationship, for instance by facilitating dialogue with the divine.

But could AI be a Thou? In contrast to Levinas, who did not consider nonhuman others except God, Buber might be more open to seeing the I-Thou relationship in nonanthropocentric ways: what the "I" or "Thou" is, is left more open at least. The emphasis is on the relation, not on the relata. This leaves some room for imagining machines taking on one of these roles rather than rejecting the possibility a priori, even if this might seem unlikely, impossible, or unacceptable to most of us. Furthermore, the initial criticism does not take seriously the possibility that people can have a (quasi-)religious experience and perpetuate religious patterns of practice when interacting with chatbots, whether for explicit and intended religious purposes or not. Experiences of technology are also dynamic. They change. While initially AI might be framed as a kind of servant to us, it can also become a lord and master that takes over control and asks things from us (e.g., more time and more data). And having started as a relationship between a human master and a friendly servant, the human-AI relationship might develop into what might be experienced as friendship or partnership. The unconditional support and friendly advice given by the AI might be *perceived* as signifying true friendship or even love. The AI is always friendly, doesn't judge us, and seems to support us. It is always there for us, available to listen to our concerns and requests. What more can one expect from a true friend? Or AI might appear as an Absolute Other, an infinite alterity that is all-knowing and all-powerful and which we can never understand. We might feel that we have to do something, respond to the ethical call that emanates from our meeting with it. For example, someone might think that AI orders them to build superintelligent machines and help it to control or even replace humanity. Or someone might come to believe that an AI system has become conscious and feel that we need to care about its "welfare."[14]

As these examples show, these experiences are not unproblematic. They may lead to attachment to technology, and they may be seen as involving dangerous illusions, potentially leading to morally problematic actions, perhaps even suicide or murder. Our intimacy with machines may also be abused to manipulate us. It may create *more* anxiety, depression, and anger instead of alleviating them. And it seems right that these technologies offer no real love or friendship. When I suggested that AI is a true friend, this contained a lot of irony. AI cannot provide what human friends can. Furthermore, Kathleen Richardson rightly argues that we cannot really have a Buberian I-Thou relationship with robots since the basis of how both humans and robots are modeled in robotics science remains an instrumental I-It relation.[15] This also seems to hold for AI. Using the language of relationships with regard to these technologies is then indeed misleading. Seeing AI systems as a relational other is not only mistaken in virtue of their properties, it also reflects back on how we view humans and human relations. To see an AI chatbot as a friend risks being on the other side of a reductionist, instrumental view of humans and human friendship.

But that does not mean that AI cannot play a role in prayer and other religious experiences. Dismissing or judging all techno-religious experiences before even analyzing them does justice neither to the role technology plays today nor to the ways religious experience continues and transforms itself today. Such changes in religious experience also happen through technology and often in ways that are not intended by designers or users of the technology. It may be that AI enables prayer in a way that occasionally leads to a genuine and beneficial religious experience, even if this is not necessarily the case (but the same can be said about using a religious book and other religious technologies and spiritual techniques) and even if AI cannot really take up the role of a Thou.

Indeed, acknowledging the possibility of religious experience through AI does not mean that we should not be skeptical of claims that AI can offer friendship or love. On the contrary. Connected to the religious need to pray, that is, the need to communicate and dialogue with the divine entity with whom one has a relationship, is the deeply human and existential need for companionship, community, friendship, and love. It is well known by ethicists of social robotics that this is precisely what these machines pretend to cater to but—so it has been argued—really fail to deliver. Psychologist and MIT professor Sherry Turkle has famously argued that while robots and AI might appear to offer companionship, they offer only a performance of companionship, an illusion that falls short of giving us true emotional reciprocity—perhaps of the kind that Buber imagined.

In *Alone Together* and *Reclaiming Conversation*, Turkle argues that we wrongly see loneliness as a problem that technology should solve.[16] We design companion robots that talk to us, but this gives us only the illusion of companionship. Technology renders real conversation and authentic communication impossible; instead, we avoid it and are busy with our phones. People form attachments to robots, to their phones, and to chatbots and other LLM-powered apps and devices. Similarly, robot ethicists such as Noel and Amanda Sharkey and Robert Sparrow have warned against what they see as the illusion of companionship and the exploitation of the human tendency to project emotions onto machines.[17] Machines lack true understanding, and vulnerable people, such as young children or elderly people on whom these robots are tested, are deceived. Similar arguments can be made about AI-powered chatbots, which can give only the illusion of companionship and which lack true understanding and capacity for friendship and love. We need conversation, for sure, but we need conversation with other humans and perhaps with a divine Thou. Amanda Lagerkvist argues that chatbots fail to meet our existential needs: we yearn for "a *someone*" who can respond, but nobody is at home. Or as she puts it in Buberian terminology, "an I is yearning for a Thou in communication with the chatbot." Yet the chatbot does not offer a Thou—neither a divine nor a mortal one—and thereby fails that "existential test." But this is not only an ethical problem. It also tells us something about ourselves and about communication: it "reveals our longing for the other, for each other," which is what is at stake in human communication.[18] Or to return to the main argument of this book using Hefner's mirror metaphor mentioned earlier: a technology like AI mirrors our existential needs and aspirations, and our craving for communication and companionship is one of them.

It is thus good to be skeptical of claims that AI can be a thou or Thou. At the same time, the performances and interactions that happen in prayer with and to AI devices and robots are increasingly similar to those that happen when humans say they communicate and interact with the divine or when they talk with a mediator, typically a priest, who assists that communication and interaction. This could potentially, but not necessarily, lead to genuine and authentic religious experiences, regardless of the "real" nature of the chatbot or *praybot.* It may well be that these machines cannot create real connection between humans and with the divine, and that they themselves lack personhood or a soul. But they might nevertheless assist in prayer and help to bring about real religious experiences.

Sometimes, and arguably increasingly, we simply don't know who is at the other end of the line or, rather, who is at the other end of the screen. Are we talking to a human or a machine, or even a god? If we don't know, and if we don't have a quick and easy way to find out, I repeat my recommendation voiced in earlier work on a relational approach to moral standing.[19] In case of doubt, we must reduce moral risk and take a precautionary approach, in the sense that we had better treat the entity *as if* it were an entity with moral standing, regardless of what we believe the properties of the entity to be. Here this means that we should communicate and treat the entity *as if* it were an entity with whom we could potentially have a relationship. We thus must respect the entity and communicative with it *as if* it were a relational other. In the context of this discussion this could mean, more precisely, that in (still exceptional) cases of doubt about the status of the entity, one should be open to the possibility of the emergence of companionship or religious experiences in communication with that unknown entity, without having (to have) certainty about its standing or properties.

As I have argued elsewhere, what matters here ethically speaking and in practice is not only, and perhaps even not so much, what the entity "is," but rather the experience and (co)performances of the humans involved, which are real and which matter, normatively and otherwise.[20] The same could be said for what matters communicatively and religiously. One could argue that people's experiences and performances matter, not the properties of the entity involved. People may pray via a robot or AI and, through this (co)performance or ritual, have religious experiences. Whether this can still be classified as illusion then depends on your view of whether prayer and its religious experiences are *themselves* a matter of deception and emotional exploitation of all-too-human aspirations (and it even further depends on your view of religion or of a particular religion such as Christianity), but not on the status of the AI chatbot or prayer robot as such. A particular AI-assisted religious experience can be real, according to the evaluation and criteria offered by a particular religion, even if people talk with a "mere" machine.

Moreover, such practices and experiences with or via AI are not religiously neutral but might change the way people think of prayer and indeed their relationship with God. Religion is not a fixed thing but variable, and technologies and rituals are not just cases of "applied religion" but shape and reshape the very meaning of religion and prayer. The technology of the book, for instance, also changed how people relate to the divine. The monotheistic religions became religions of the Book and text became essential as a medium of religious communication with and about God. Praying became more a matter of reading and writing than of oral communication. AI and other digital technologies are most likely already transforming these (re)connections (*religions*) and communications to the divine, even if it may not yet be entirely clear how. As I said earlier, technologies are never just means; they have unintended consequences and even change the very practices and goals they were meant to assist and mediate. AI is not only shaped by religious ways of thinking, in the sense that it is used for religious purposes. Indeed, it is likely that religious experience is also shaped (and will continue to be shaped) by AI. Rather than passing a quick moral judgment on the use of AI in prayer, or any other technological practice and phenomenon for that matter, it is important to first recognize the possibility of these unintended influences and transformations and study them; one can then evaluate and debate whether they are good or not.

This discussion about the more-than-instrumental role of technology can be compared with one that is as ancient as Plato's writings: the question of whether *writing* itself is a good technological invention. According to Plato, it is not. In the *Phaedrus*, he argued that writing stimulates forgetfulness in the reader.[21] It produces only the appearance of knowledge. Today the same is said about chatbots and AI. But if they are somehow fake, what is real? If they are the surface, where is the depth, the origin of the message in the communication? How do we get to the truth? Platonic thinkers point to the value of orality, the original message, the spoken word (logos).[22] They fear that it risks getting lost in a (Western) culture dominated by writing and text. Interestingly, as I already suggested, for a very long time in religious and cultural history, reading (including the reading of religious texts) was an oral activity. People did not pray by means of reading but by speaking or singing. Often praying was also a communal activity. Reading silently was rare and emerged much later. For example, in his *Confessions*, Augustine remarked that Ambrose often read silently.[23] This perplexed him. In the ancient world of Greece and Rome, reading was done aloud, and memorization was seen as important. Silent reading became more common only with the invention of the printing press in the fifteenth century, which was also the beginning of more individualist and introspective approaches to reading and thinking—in religion and elsewhere in intellectual life. Prayers, however, often continued and continue to be performed by means of recitation, chanting, and singing. Consider for instance the Lord's Prayer in Christianity, which Christians are supposed to have memorized, and which is recited or sung during mass.

Media scholar Walter Ong wrote in *Orality and Literacy* about how oral praying traditions were fundamental in early societies and how literacy changed the nature of communication and cognition.[24] McLuhan noted similar shifts when he discussed the impact of the printing press in *The Gutenberg Galaxy*.[25] Orality is often seen as having its own unique quality. Nevertheless, the technology of writing took over. In the course of cultural and religious history, praying became increasingly a matter of reading (and writing) text. This culture of literacy with its new technologies led to a focus on books and texts in Western religions. As I said, it turned the monotheistic religions into religions of the Book. In contemporary times, the culture of literacy has enabled the use of contemporary text-based technologies such as websites and LLMs, which are trained on texts from the web. As today's most successful forms of AI, LLMs are thus exponents of the triumph of the culture of literacy. Even if they also include some forms of orality such as chatting—forms that Ong has called "secondary orality"—they are firmly anchored in the processing of text.[26] And all digital technologies are based on written code.

Often the development toward praying in the form of reading and writing is seen as problematic for religious experience. If religion is all about relationship and connection, is this not better achieved with oral and communal practice? In his *Confessions*, Augustine argued in favor of chanting and singing, which according to him facilitate a deeper connection with God; such praying, he argued, has more emotional and spiritual impact. He described how hymns moved him to tears and he emphasized the communal aspect of hymns and psalms; singing integrates body, soul, and community. People with intellectual abilities may read religious texts and, in this way, get closer to God. But, Augustine argued, singing also helps to get weaker minds into a devotional mood. Yet he also warned that the singing should not distract from the message. It should pour (religious) truth into the heart.[27]

With his defense of orality, Augustine stands in the Platonic tradition of what Jacques Derrida called "phonocentrism": in spite of the obvious success of literacy culture, in Western thought speech (voice) is still seen as more authentic than writing (text).[28] Logos is linked to voice. The spoken word is seen as being closer to the origin of the message that is being communicated (the speaker and what the speaker had in mind before speaking) and closer to the truth—here, closer to God and His message. There is less mediation, if any. Writing, one could argue in a Platonic fashion, is too much about appearances. It is only a shadow of the real. There is too much distance from the truth. It leads to detached thinking and communication. Similarly, LLMs can be said to merely imitate; authenticity is lost, the communication is too fake. What LLMs produce is removed from the oral source and even from the original texts. There is a kind of double alienation in this sense. What we need, in prayer and elsewhere, is oral forms of communication, which are more authentic and bring us closer to the divine Logos.

Against these criticisms of prayer mediated by writing, and by LLMs, one could argue that while the nature of prayer and thinking has indeed changed—generally there is a loss of the communal and voice aspect, even if some oral practices such as singing together persist—printing and texts have not harmed Christian or Jewish spirituality and religion but instead further supported their development and reach. Similarly, one could argue that while there is an aspect of orality to chatbots (but orality with the absence of a human voice; the "voice" is synthetic, and therefore one could object that it is a "fake" voice and hence "fake" orality, a mere appearance of orality), chatbots trained on Christian texts and text-based AI priests continue the literacy tradition rather than a culture of orality, and that this has similar mixed effects. Literacy in the form of AI-based devices continues the development of a more individual and text-mediated relationship to God. One might worry that this may distance people from their lived relationship with God. At the same time, one could argue that the technologies provide a means to potentially bridge that distance, in so far as they still, or nevertheless, offer a communicative bridge to the divine. Seen from this perspective, AI and its link to a culture of writing is not necessarily bad for religion. As a means of religious communication, that is, as an instrument that aims to *re-ligare* (Latin for "reconnect," "bind again"), AI might help people to reconnect to God, to the texts and authors of their religious tradition, and to each other.

An evaluation of the religious significance and potential benefits of AI thus at least partly needs to reflect the ambiguities and difficulties raised by that older, arguably more fundamental medium and technology that still shapes much of our current practices, including conversation and prayer: writing. When people speak to God through contemporary technologies, writing is the uninvited third conversation partner. Even if the communication is through voice (human and synthetic), writing is present in and through LLM technology. And this is also the case for those who pray. In line with Derrida, one could argue that even if people do not explicitly use texts, their oral praying is shaped by the medium of text and writing. Or to say it in a McLuhanian fashion: the medium is the prayer.[29] In so far as AI remediates writing and is involved in prayer, praying is also always a bit of reading and writing texts—with all its advantages and drawbacks.

6
Foretell the Future: Oracles, Prophets, Saviors, and the AI Apocalypse

Since ancient times we have wanted to know what the future holds for us. In ancient Greece, oracles would foretell the future. A famous oracle and one of the most important religious centers of ancient Greek society was that of Delphi. The high priestess of the temple of Apollo, the Pythia, was famous throughout the ancient world for divining the future and giving advice to powerful people. Developed in the eighth century BCE if not earlier, it was the most prestigious and authoritative oracle, mentioned by authors such as Plato, Aristotle, Plutarch, Euripides, and Sophocles. Apollo's temple was located on the sacred slopes of Mount Parnassus in central Greece, where according to legend Apollo had slain the snake Python. Both kings and common people sought the advice of the Pythia, who in turn received gifts. First the petitioners had to undergo purification rituals and would bring donations. Then they could pose their questions to the mistress of Apollo, who had the power to enter in ecstatic union with the god. The Pythia would undergo sacred rituals such as fasting, bathing in the Castalian Spring, drinking holy water, and inhaling gases emitted from geological fissures. A goat would be sacrificed and its organs examined and burned. Seated on a tripod chair, the Pythia would fall in a trance and—at least according to the most popular story—mutter cryptic, if not incomprehensible words. She was considered to be under divine possession (*enthusiasmos*, *ἐνθουσιασμός*) by Apollo. The Apollo priests, who had not only religious but political and diplomatic knowledge, would translate the Pythia's statements into ambiguous interpretations that they could use to their advantage. They thus became powerful—a power that would wane only with the rise of Christianity.[1]

That interpreters of divine messages gain power is not an exception in religious history. Bertrand Russell, when commenting on the gap between the founders of religions and churches, put it as follows:

> As soon as absolute truth is supposed to be contained in the sayings of a certain man, there is a body of experts to interpret his sayings, and these experts infallibly acquire power, since they hold the key to truth. Like any other privilege caste, they use their power for their own advantage. They are, however, in one respect worse than any other privileged caste, since it is their business to expound an unchanging truth, revealed once for all in utter perfection, so they become necessarily opponents of all intellectual and moral progress.[2]

Russell mainly commented on the monotheistic religions, which demand submission to a revealed truth; he was writing about Christianity in this excerpt. He claimed that religion is based on fear (see the discussion of Hobbes in chapter 2) and despotism. Science can help us to get over our fears and liberate ourselves; we should cultivate a free intelligence. But in this context his comment points us to the political importance and power of religious mediators, in Delphi but also in Rome, Jerusalem, and elsewhere. The messages of the oracle of Delphi, with the Pythia being seen as a mediator of the god Apollo, were in turn interpreted and channeled by the Apollo priests. This then influenced people in politics, business, and even philosophy, leading to further interpretations and power games.

Interestingly, one of the most well-known people from antiquity who is said to have been impacted by the oracle of Delphi was Socrates. The oracle's prophesy profoundly influenced not only his life but also his philosophical project—and thereby Western philosophy and culture at large. As is recounted in Plato's *Apology*, the priestess famously said that there was no one wiser than Socrates. This statement set Socrates on a path to question people in Athens who had the reputation of being wise, but he found that they lacked true understanding. Socrates concluded that he at least recognized his own ignorance, that he knew the limits of his knowledge. The interaction with the oracle, together with the famous Delphic principle "Know thyself," thus helped him develop his Socratic method and his views on knowledge and ethics. Socrates was grateful for this. At the end of his life, during his trial for impiety and corruption of the young, Socrates paid tribute to the god Apollo. He said that Apollo commanded him to take on the role of the gadfly, the annoying, stinging questioner of those in the city who thought they were wise.[3] It was politically dangerous and Socrates died for it.

Today we have AI oracles, which—through incomprehensible operations and using the *pharmakon* (drug) of advanced technology—predict our future and mediate our relation to what is beyond our control. Prompt engineers, developers of chatbot apps, data scientists, and AI consultants are the new priests and priestesses who help us to predict our future and try to make profits from it. AI itself becomes an oracle that is eagerly consulted by judges, police investigators, researchers, and finance specialists. And ultimately by everyone. LLMs are used everywhere. Even in the modern age, we want fortune tellers because prediction is power. This time we aren't using goats or consuming drugs; the prediction is based on data and statistics. Social welfare institutions and insurance companies, for example, try to predict people's future behavior by using AI. Based on browser history, social media activity, location data, and biometrical data, AI-driven apps predict consumer behavior. Consumers, on their part, get solicited and unsolicited recommendations by AI apps about their health, relationships, everything.

Generally, we welcome AI and its predictions. We think we know more. We think we learn something about ourselves. Perhaps we even have the illusion that we can know everything now. But we lack Socratic wisdom. We forget the Delphic principle. We forget the limits of our knowledge. Today we have so much more information via our large language models, and we have insights and opinions from all over the world via AI-powered social media, but we lack true understanding. We fail to know ourselves, personally but also collectively. We do not understand that the future will always resist control and colonization, that the *terra futura* will always remain beyond our greedy reach, that we cannot fully tame future uncertainty, change, and contingency. And we certainly fail to know what our own technologies do to us. As Helga Nowotny argues in her book *In AI We Trust: Power, Illusion and Control of Predictive Algorithms*, we need wisdom that acknowledges the limitations of AI.[4] AI gives us only the illusion of control. We want predictive certainty and control but the future is full of uncertainties and unknowns. We blindly trust AI, failing to exercise our own imagination, and so we do not see what the predictions do to us. As AI's prophesies become self-fulfilling, we close the future's open horizons.[5]

Moreover, like the Greek oracle and the prophets in biblical times, today human AI prophets predict the future of technology and indeed the future of humanity. Popular authors such as Ray Kurzweil, Nick Bostrom, and Yuval Harari are the new prophets who pretend to know what the technological future holds for us. Building on the predictions of earlier techno-prophets such as Gordon Moore, John von Neumann, I. J. Good, and

Hans Moravec, who have already speculated about the accelerating pace of machine intelligence, Kurzweil and Bostrom argue that computing power follows an exponential curve and that this will result in superintelligent AI. According to Kurzweil, this will eventually lead to a merging with humans and digital immortality. There is no need to be content with your mortal body and accept death. Harari also thinks that we will evolve into a new species. We will go out into the cosmos and colonize new worlds. Before that happens, there will be a centralization of power in the hands of those who control data, and there will be more inequality. Humanism will end since algorithms will understand us better than we understand ourselves. And perhaps humanity as such will end. Bostrom and so-called longtermists warn about the long-term risks of AI. If we cannot control AI, if AI's goals are not aligned with those of humanity, and if AI systems evolve in unpredictable ways, it could cause human extinction. In any case, all three agree that AI will cause a fundamental transformation of society and humanity.[6]

As prophets of the technological future, these authors are not only influential high priests in what seems an essentially colonial and puritanical religious movement and messianic cult aimed at technological salvation, as Epstein has argued, or heirs of Russian cosmism with its beliefs that death can be overcome and that humanity's future lies in outer space, or missionaries of a form of technological Gnosticism that exhibits contempt for the human body.[7] They also continue the Jewish traditions of prophecy. In ancient times, prophets (*nevi'im*) played a crucial role as intermediaries between God and the people of Israel. As divine messengers, they communicated God's intentions to the people. They had a central role in ancient society. Prophets such as Isaiah, Jeremiah, and Ezekiel addressed religious but also moral and political issues. They predicted and warned. For example, Jeremiah said that because the people of Judah had turned away from God and had failed to uphold justice, Jerusalem would be destroyed and fall to the Babylonians; a period of exile would follow. Considered chosen by God to deliver His message with corresponding authority, the prophets often advised kings and military leaders. Nathan, for example, was a trusted advisor and friend to King David. Prophets were revered figures and their messages could provide divine legitimacy to rulers.

Yet situated outside established religious and political structures, the Jewish prophets also criticized those in power and their messages could challenge the actions of rulers. Drawing on the values of their religious tradition and appealing to justice, they spoke truth to power. Sometimes that turned out badly for them and they were persecuted. In one such case Jeremiah was beaten and imprisoned because of his message of doom for Jerusalem, and Elijah's life was endangered when Queen Jezebel pursued him after his opposition to the royal cult of Baal. To be a prophet was risky. But their work influenced Jewish theology, they are revered in Jewish culture as models of courage, and their prophesies became part of the Christian Bible.

Today people such as Harari advise the CEOs of Big Tech companies, the new kings who want to know the technological future. Business leaders and politicians also want to know the future of society. How will the markets change? Where should they invest? What is the technology of the future? What does AI mean for our nation and for our administration? But the ambitions and expectations reach further, even into the big questions concerning the human condition and the future of humankind. Inspired by transhumanism, a movement that argues that humans should use technology to enhance the human condition beyond current biological constraints, some people now entertain the existential and (quasi-)religious expectation that the new prophets and new technologies such as AI will help to liberate humankind from its self-chosen and unnecessary human and mortal limitations. With new technologies, we now have the freedom and opportunity to change the human condition. Transhumanist prophets instill the hope that the new technologies will help to overcome the human, which becomes obsolete. Current humans have to be improved or, better, transcended and replaced entirely by artificial intelligence and related technologies. The future doesn't need us, as Bill Joy put it in a famous *Wired* article when he warned that new technologies threaten to make humans obsolete.[8] Some AI prophets may also support the ultra-libertarian or even authoritarian ambitions of

Big Tech companies and their leaders. Through their influence, they do more than prediction: they also help to *make* the future. They reveal the future (the word *apocalypse* originally meant "revelation" in Greek) but are at the same time counted among its architects. Like the Greek priests and Jewish prophets, they also assume some political power through their influential teachings and prophesies. As experts who reveal and interpret the truth about AI and its future, they may not have formal political positions, but they do have soft power. They shape opinions, influence decision-making, and set policy agendas. While in principle they could also fulfill a critical role (and to some extent they do), they mainly legitimize those in power: leaders in Big Tech, national governments, and international organizations. Harari has been a regular guest at the famous World Economic Forum in Davos, Switzerland, and has been welcomed by President Emmanuel Macron at the Palais de l'Élysée in Paris, France. Bostrom similarly has influenced tech leaders such as Elon Musk, Mark Zuckerberg, and Bill Gates.

Their imaginings of the future follow a well-known religious pattern and are all about salvation and endtimes. They are about expectations regarding the end of the present age or the end of the world, the final destiny of humankind (eschatology), and the coming of a new order. They are apocalyptical. The ancient prophets predicted a new divine and earthly order after a period of judgment. Their prophesies were about radical political and spiritual transformation. In the Hebrew Bible we find the Book of Daniel, which predicts a future of drastic cosmic and political change. The prophesies talk about mad kings, monstrous beasts arising from the sea, and angels that interpret visions. The Book of Daniel predicts the rise of knowledge, talks about the "image of the beast," and speaks of a future kingdom that will crush all others. One of the beasts is described as having "iron teeth" in a passage that is sometimes interpreted as saying something about the technological future and power: "In my vision at night I looked, and there before me was a fourth beast—terrifying and frightening and very powerful. It had large iron teeth; it crushed and devoured its victims and trampled underfoot whatever was left. It was different from all the former beasts, and it had ten horns."[9]

The prophesies also contain a prediction that has become very influential in the course of Western religious history. In Daniel 9 it is said that a "messiah"—from the Hebrew word for "anointed" (*mashiach*, מָשִׁיחַ)—will come and that this anointed one, also called "the ruler" or "leader," "will be put to death."[10] In ancient Israel, people who were chosen for the role of king, priest, or prophet were anointed with oil. In the Jewish religion, Jesus is a prophet, a messenger of God. But according to Christians he is a very special one, a divine revelation in person and the Son of God himself. He is seen as the foretold messiah, causing political turmoil and enabling radical spiritual renewal in the form of Christian salvation. This can be understood in terms of personal spiritual salvation, but it can also be interpreted as collective, earthly, and political. In spite of his answer to Pontius Pilate during his trial that "my kingdom is not of this world," Jesus was seen not only as a spiritual leader but also as a political rebel and (potential) liberator—a "King of the Jews," who not only disrupted activities in the temple in Jerusalem, itself a political act, but, after his entry in Jerusalem as "Son of David," was also expected to free the Jews from Roman rule and establish a Jewish kingdom.[11] Crucifixion was the punishment for political rebels and criminals.

Another relevant biblical book is Revelation (or the Book of the Apocalypse), initially attributed to the apostle John and the final book of the New Testament. It tells a story that involves the infamous Four Horsemen—conquest, war, famine, and death—but also the Seven Angels that unleash plagues and disasters, a dragon that pursues a child (Christ), the Fall of Babylon, the defeat of the Beast, and the vision of the New Jerusalem. The book's narrative structure can also be found in many Hollywood movies: After setting the stage and introducing the protagonist, there are problems and challenges for the hero, there is suffering, and then there is a happy ending and the offer of hope. There are a number of battles; there is crisis, the stakes are raised; but in the final epic battle in which the rebels confront the empire, the forces of evil are defeated, leading to heaven on earth.

An influential interpretation of the Revelation story is that now there is suffering and conflict, but eventually good will win. We just have to wait, be good, fight evil, and in the end all will be fine. This way of thinking has been used to legitimize wars (the enemy is serving Satan and needs to be defeated, after which all will be good) and to support existing unjust political order. Today you may be in a bad social position, you may be a slave, but if you are good and have faith, there will be a bright future. This world is horrible but the next one will be fine.

Marx famously critiqued the idea of passively enduring suffering for a future utopia, and I have mentioned Nietzsche's criticism of religion as an instrument for social control and conformity. Similarly, the Revelation narrative can be seen as providing opium for the people, encouraging them to endure suffering and accept the system instead of actively dismantling oppressive structures. At the same time, the story can also be interpreted as a criticism of oppressive political systems. Revelation can be read as a Jewish criticism of persecution under Roman rule. Refusing the mark of the Beast is an act of resistance; the New Jerusalem is then a vision of a just society without oppression. This interpretation thus highlights the subversive and liberatory potential of the story. We can and should work toward ending injustice. That being said, the suggestion that the collapse of oppressive regimes is inevitable can again lead to passivity. If capitalism and empire will fall anyway, then why not just wait for that to happen instead of revolting?

In medieval times and afterward, end-times have regularly been predicted by figures such as Hildegard of

Bingen, Martin Luther, and Nostradamus. Here, too, the content of the predictions were a mix of religion and politics. For example, Hildegard of Bingen spoke out against abuses of power and called for reform. Luther saw in the pope and the Turks the coming of the Antichrist. Some interpret Nostradamus as having predicted the French Revolution, the rise of Napoleon Bonaparte, the rise of Hitler, and current and future wars in the Middle East. And following the Book of Revelation, some Christians, both in the religion's early history as well as in today's United States for instance, believe in millennialism, which predicts that before the Last Judgment, there will be a thousand-year Messianic Age in which Christ will reign on earth. Revelation 20:1–6 speaks of an angel who binds Satan and describes a millennium in which Christ reigns with his saints. Millennialists wait for Christ's return on earth and the establishment of his kingdom, literally or spiritually.

The medieval Joachite tradition is another example of apocalyptic and eschatological thinking with, again, a political dimension. Named after Joachim of Fiore, an Italian theologian who founded a monastic order, it held the belief that there will be a new age (the Age of the Spirit) when a spiritual elite will lead humanity to a purer form of spirituality—replacing the corrupt Church. This inspired some radical Franciscans and Protestants, but is also echoed in utopian Marxism. Indeed, Marx has been accused of propagating a secular version of millenialist and apocalyptic thinking: he has been interpreted as having predicted that a revolution will necessarily result in the end of capitalism and bring with it the utopia of communism. Again, a decisive break with the past is seen as (necessarily) leading to a radically different future. Ernst Bloch, for example, has argued in *The Principle of Hope* that modern utopian socialism, with its hope for a better world, continues apocalyptic and millennialist thinking—including Joachite ideas.[12]

A related religious concept and phenomenon is doomsday cults. These are religious groups that believe in apocalypticism and millennialism. They predict imminent catastrophic events and sometimes the end of the world, for which they prepare. Doomsday cults typically appear in times of economic and political instability and anxiety. Charismatic leaders claim to know the future and exercise power over their followers. Historical examples include the Taborites in fifteenth-century Bohemia, the Münster Anabaptists, who held beliefs that resemble Joachite millenialist thinking and who took over the German city for a while in the sixteenth century, and doomsday sects in the United States and Japan in the 1990s. Often these cults have ended tragically. For example, the Anabaptists' rebellion was violently suppressed, the California group Heaven's Gate ended in mass suicide, and in 1995 members of a Japanese doomsday cult carried out a gas attack on a Tokyo subway. But their political impact is not limited to radicalization and terrorism. The apocalyptic beliefs of doomsday cults can also shape public discourse about technological risks and contribute to the rise and maintenance of societal fears and anxieties. Beyond specific cults, Christian apocalyptic thinking and a fascination with end-times and prophecies have always been a part of modern American culture and have impacted public life.[13] Perhaps apocalyptic and doomsday thinking are even a fixed ingredient of modern Western culture, including Western thinking about modern technology. Consider for instance the end-of-the-world thinking about nuclear technology that captured the public imagination in the second half of the past century: many people were anxiously awaiting the bomb, that is, total annihilation by nuclear war. As Derrida argued, nuclear war was a matter of not just technology but also rhetoric, discourse, and persuasion.[14] A matter of bombs, but also a matter of texts. And that rhetoric and those texts were undoubtedly apocalyptic.

This is also the case today: apocalyptic and doomsday beliefs about AI circulate and shape public discourse and political imagination about the technology and its risks. AI is about technology but also about rhetoric—and that rhetoric is apocalyptic. Robert Geraci has proposed the term "Apocalyptic AI," which suggests that apocalyptic thinking is not something of the ancient past but is alive and well today and present in popular science books on AI and robotics.[15] In a similar vein, Noreen Herzfeld calls AI "a millenarian project," allowing people to project "eschatological hopes and fears" onto AI.[16] Salvation or destruction: it is now all in the hands of AI. It is yet another example of how, in Geraci's words, "our techno-scientific heritage is grounded in the religious life of the Western world" and its ancient sources.[17]

What are the apocalyptic predictions regarding AI? What, in particular, is the transhumanist eschatology that tends to be so influential in contemporary narratives and discussions about the technological future? AI is said to lead to an end-time: there will be a Singularity, when everything will be different.[18] There will be an uncontrollable intelligence explosion, resulting in superintelligence. The outcome of that development is expected by Bostrom to be extremely good or extremely bad—but extreme in any case.[19] Societies will be fundamentally transformed. To use a metaphor from Matthew, "not one stone will be left on another."[20] The current development of AI is understood as preparing for that future scenario of radical change and disruption. It is predicted that the current technologies will be superseded by artificial general intelligence (AGI). AGI is a type of AI that will first match and then surpass our human capabilities. In other words, one day our machines will be superintelligent.[21] Some see current LLMs as steps toward emerging AGI, whereas others think we need a different kind of AI. But most Singularity believers agree that the end-time is not so far away. The end (or, one might say, the new beginning) is near. According to Kurzweil, "the Singularity is near"; in his book with this title, he has predicted the "profound change" for 2045.[22] Most other predictions, such as those by Moravec or even Musk, also say the mid-twenty-first century, although Bostrom keeps it vaguer, drawing the line sometime in the twenty-first century.

What happens then? It is believed that the consequences of superintelligence for human civilization will be radical and irreversible. When superintelligent AI takes over, humanity as we know it will end. We will become transhuman beings—significantly technologically enhanced—or merge into superintelligent machines, spreading out into the cosmos, where our "mind children," as Moravec put it, will flourish: our offspring in the form of superintelligent AI.[23] Human intelligence will be nothing compared to the superintelligence of the new beings living in space in the future. Perhaps those beings will be digital. It is imagined that billions of digital people—or, rather, superintelligent entities—will live in computer simulations.

Influenced by the work of Bostrom but also that of Toby Ord, William MacAskill, and Eliezer Yudkowsky, some so-called longtermists argue that we should prioritize the long-term, far-future interests of those superintelligent beings because the moral weight of them, our future "potential," is enormous.[24] From this longtermist and utilitarian point of view, the interests of the few people living today do not count for much. And it's unlikely that we have a choice anyway. Humanity is doomed. It is obsolete. The end is near and it is also humanity's end. In a *Time* piece, Yudkowsky lays out his expectations clearly and unambiguously when he writes, "the most likely result of building a superhumanly smart AI, under anything remotely like the current circumstances, is that literally everyone on Earth will die."[25] As Émile Torres, a prominent critic of longtermism, argues, this kind of thinking is dangerous. If the greater cosmic good needs to be prioritized over the present and near-future interests of ordinary humans, then current poverty, climate change, and other problems of present and near-future generations don't matter that much.[26] They don't matter in the grand scheme of things.[27] And doomsday thinkers, from medieval times to today, are only concerned with the grand scheme of things. They do not care about you and me now; we are supposed to die while the new, more important future is being prepared.

If this were just the marginal, sectarian, and esoteric worldview it seems to be, then there would be no problem with regard to the discourse on AI, and indeed the politics of AI; it could simply be ignored. But what started in the fringes of elite British universities and Silicon Valley research institutes has today become the faith and ideology of powerful Big Tech CEOs such as Musk and Sam Altman. Musk not only sponsors transhumanist and longtermist centers such as the Future of Life Institute and, earlier, Bostrom's Future of Humanity Institute, but also integrates the ideology in his visions of, and plans for, the technological future. For Musk, colonizing Mars is a stepping stone in the new, longtermist story of human evolution and civilization. Current AI prepares the way for our Brave New Future. Big Tech will help make that future happen. This may not be good news for humanity. It may well be that Musk wants to save part of humanity (perhaps himself and some of his billionaire friends), but only to colonize space, spread out into the cosmos, and thus participate in the new future in the form of enhanced beings or simulations. Most current mortals are not needed; humanity will go extinct anyway.

Following his warnings about future AI, Musk has been linked to the so-called "AI safety movement" that trades in superintelligence expectations and hypes AI while at the same time offering doom scenarios about it. Backed by billionaires such as crypto magnate Sam Bankman-Fried and entrepreneur Dustin Moskovitz, the movement has tried to influence, if not colonize, the field of AI ethics: it only wants to talk about AI risks in the light of its longtermist vision. Politically, it has even pushed in an open letter to pause AI research based on the doomsday prediction that sufficiently advanced AI can lead to human extinction.[28] Signatories included Musk, Harari, and machine learning researcher Max Tegmark, but many others in the AI community are drawn into this. AI has thus increasingly become a doomsday cult, and one backed by dollars. Billionaire tech entrepreneur Peter Thiel is also influenced by what Torres calls an "intoxicating new secular religion."[29] Prepping for the longtermist apocalyptic future, Thiel has a retreat in New Zealand. In case of civilizational collapse he plans to hide there with Altman.[30] The latter shares Musk's belief in the apocalyptic potential of AI. Like Musk, Thiel, and other tech billionaires, Altman has engaged in doom thinking about AI and its "existential risks." As Torres points out, that term is misleading. As far as they are longtermists, these key players in the AI world, who have significant global impact, do not worry much about the future of presently living humans such as you and me. At most, they care about the far future and are only interested in current problems if there are existential risks today that threaten the realization of that future. What happens today is a mere ripple in a much larger apocalyptic narrative and salvation history in which these Big Tech leaders see themselves as playing a key role.

The latter is unsurprising from the point of view of religious history. Prediction is power. The prophet can become the savior. The king who heeds the oracle's warnings can gain a lot. Listening to messengers of the future like Harari and Bostrom who predict AI's longtermist salvation history, Big Tech CEOs such as Musk and Altman can easily present themselves as saviors who not only know what will happen to humanity but also promise to save us from the evils of ordinary humanity and mortality, leading us into the new times of Singularity apocalypse and superintelligence. They can appear to the general public as all-knowing and soon all-powerful guides in the Brave New World that is coming. This gives them political power. And the same power, to some extent at least, accrues to their advisers, even if their message is not always well received. Like the Pythia and especially like Jesus and other biblical prophets, authors such as Harari might become political and revolutionary willy-nilly, even if they claim to "merely" speak from and for the inevitable techno-cosmic future.

We can already see that the tech billionaires are ready to assume more political power. In Musk's case this has become very concrete. He used his social media platform X, formerly Twitter, to gain political power, and was appointed as head of the "Department of Government Efficiency" (DOGE) in Trump's administration with access to the US Treasury's payment systems. Before the 2024 US elections, Musk openly declared that he wants more formal political power and that he is "willing to serve," as he made clear in an interview.[31] A similar readiness to serve may well be found among the prophets of AI's already determined future, who whisper their predictions of glory and doom in the tech king's willing ears.

Attending to the history of religious ideas thus helps to answer a pertinent question ironically posted on X: "How did outlandishly fringe AI doomsday cults gain so much unwarranted political influence?"[32] The questioner is not an eccentric leftist political theory professor but Yann LeCun, Meta's chief AI scientist.

7
Let Me Live Forever: Gilgamesh, Resurrection, and Immortality Machines

When his friend Enkidu dies, King Gilgamesh, king of Uruk in ancient Mesopotamia and known as a powerful tyrant, is struck with grief. Fearing his own mortality, he embarks on a quest to learn the secret of eternal life. King Utnapishtim, who has survived a flood by making a ship to preserve life (in a story predating that of Noah's in the Bible), tells him that immortality is a gift of the gods and cannot be earned. Nevertheless, Gilgamesh continues his quest because he also hears from Utnapishtim about a plant that gives everlasting youth. He eventually finds the "Plant of Rejuvenation" at the bottom of the sea. But on his way back home, a snake—an ancient symbol of rebirth and transformation—steals the plant. Gilgamesh is devastated. He is forced to acknowledge his mortality. Immortality is reserved for nature or the gods; for humans, death is inevitable. Gilgamesh finds consolation in the lasting legacy of his deeds and works. His city and people will outlive him.

The Gilgamesh myth is one of the oldest known stories in human history. It is certainly not the only story that deals with the theme of human mortality and the search for immortality; many human cultures and civilizations have been preoccupied, if not obsessed by it.[1] In a sense, civilization is the *result* of what Stephen Cave has called the human "will to immortality": much of the history of religion, philosophy, and art can be seen as an outcome of our wish to perpetuate ourselves.[2] The epic of Gilgamesh reflects what is probably the most defining existential struggle of humans and humanity: Death is a scandal. We cannot and do not want to accept it. We want to live forever. Or as 1 Corinthians puts it, "The last enemy to be destroyed is death."[3]

Immortality is not always seen as the endless continuation of physical existence. In Hinduism, for example, immortality is understood as spiritual liberation of the soul (*atman*) from the cycle of birth, death, and rebirth (*samsara*). Self-realization leads to spiritual immortality, a state in which the soul is free from worldly attachments. Hindu texts such as the Bhagavad Gita and the Upanishads tell of atman, the individual soul or self, that is united with Brahman, the world soul, the supreme reality. The liberated person realizes that there is an identity between atman and Brahman: the self, in its original and pure form, is Brahman (see the famous mantra "Tat Tvam Asi," तत् त्वम् असि in Sanskrit, or "Thou Art That" in the Chandogya Upanishad). Realizing this unity is the key to immortality. As the Bhagavad Gita teaches, the soul has an eternal nature and does not disappear when the body dies: "The soul is neither born, nor does it ever die; nor having once existed, does it ever cease to be. The soul is without birth, eternal, immortal, and ageless. It is not destroyed when the body is destroyed."[4]

Many ancient Greeks also believed in the afterlife or the immortality of the soul. In the *Iliad* and the *Odyssey,* the afterlife is described as a shadowy, ghostly existence in Hades, where the dead go on to live. Later a stronger idea of immortality gained traction due to the influence of religious movements (e.g., Orphism and the Eleusinian Mysteries) and philosophy. Pythagoras had already taught that the soul is immortal and passes through a series of lives. In contrast to later philosophers such as Aristotle, who was skeptical about the idea, Plato also argued in the *Phaedo* that the soul is eternal. It exists before our life and survives the death of the body. The souls of people who have died go to another world and come back.[5] In the *Phaedrus,* Socrates said that the soul is immortal and proceeded to argue for it. He described how the soul settles down in an earthly body. The resulting living being, however, is not immortal.[6]

Nevertheless, in some ancient religions, and especially in those that would influence later Western cultures and religions, the physical body is also seen as having potential for immortality. The ancient Egyptians believed in the resurrection of the body after death: the deceased would continue to live in the afterlife, where their bodies would live on in a transformed and idealized version. The practice of mummification, an expensive process and ritual that was reserved for the wealthy and powerful, such as pharaohs and high-ranking officials, and that reinforced the elites' power and the idea that they were chosen by the gods, reflects the belief that the body needs to be preserved so that it can be reanimated in the afterlife, when the soul (*ba*) will reunite with the body (*ka*). The Osiris myth tells about his resurrection: he becomes the god of the afterlife who judges the dead. Ancient Zoroastrianism, originating in Persia and having a profound influence on Jewish thought during the Babylonian Exile, also holds a belief in the resurrection of the dead at the end of time: they will be resurrected in their physical bodies and judged. Adherents of Mithraism, a mystery religion practiced in the Roman Empire, believed in the resurrection of their savior, Mithras.

Christians also believe in resurrection, specifically two instances: the resurrection of Jesus Christ (a cornerstone of Christian theology, which, as already pointed out, has roots in Jewish eschatology and is recounted in the gospels and elaborated in Paul's writings) and a general resurrection at the end of time, when the dead will be raised from their graves and judged. It is believed that the righteous ones will be granted eternal life in a transformed, glorified body, while the wicked will be eternally separated from God (see especially John, Corinthians, and Thessalonians). The Apostolic Creed (second century CE) and the Nicene Creed (325 CE) consolidated these beliefs. Traditionally, Christians also believe in heaven and hell: heaven is a place of peace in the presence of God, an eternal communion with God, whereas hell is a state of separation and suffering. Catholics add that there is purgatory, a state of temporary purification before entering heaven. Where each soul goes is determined in a judgment. At the end of time, when there is the second coming of Christ, there will be a final judgment. Immortality is thus something for the future. First, we die. Based on the Gospels of John and Paul, however, some have argued that eternal life is also a present possibility if we believe in God. John 5:24 reads, "Very truly I tell you, whoever hears my word and believes him who sent me has eternal life and will not be judged but has crossed over from death to life." Being a Christian is then seen as having "a new life" in Christ.[7] In Luke 17:21, Jesus says that "the kingdom of God is in your midst."

Note that the Christian belief that souls will be judged and divided into the saved and the damned metaphorically mirrors ancient agricultural practices of sorting and classification and has been used to justify practices of inclusion and exclusion within society. Agricultural sorting is used as a metaphor for judgment in the New Testament (e.g., the parable of the wheat and tares) and this logic is extended not only into the spiritual realm (here eschatology) but also into the social and political realm, thus reinforcing social and political hierarchies. Souls are sorted, people are sorted. In medieval times in the West, judgments about the soul were linked to social and political control. Those who did not conform to social and religious norms were often marginalized or persecuted. As Michel Foucault argued in *Discipline and Punish* and in his work on sexuality, modern power has operated through disciplinary mechanisms such as examination and confession as tools of control and surveillance.[8] This has been to some extent a continuation of Christian practices. Submissions to judgment (in light of the possibility of salvation and immortality) and "pastoral power" shaped and governed individuals' souls based on knowledge of not only their behavior but also the inside of their minds. This led to patterns of inclusion and exclusion and reinforced the power of religious (Church) and political institutions (such as the modern state), which have continued forms of pastoral power, and perhaps try to continue it in new ways today with the help of technology.[9] With AI, both private corporations and nation states not only engage in surveillance of our behavior but also collect knowledge about, and potentially manipulate, our minds. Mind and soul are being quantified, analyzed, and manipulated.[10] Accounting of the soul is now done by digital means.

But AI is also linked to the quest for immortality. Contemporary digital technologies such as AI can be interpreted as responses to these existential and religious aspirations regarding immortality—and indeed as interacting with the related power and politics dynamics. They are what I call "immortality machines": attempts to reach immortality by technological means.

This is so in at least the following senses.

First, digital technologies provide different worlds and promise new lives. Virtual worlds and computer games, now more realistic than ever with the help of AI, promise liberation from the dull and risky mortal and bodily life. In these exciting new worlds and narratives, it seems we are no longer bound to our physical bodies or even our old selves; we can be heroes or gods, we can be who or what we are or want to be. Our digital selves, once decoupled from our physical bodies, achieve a freedom and state that seems immortal or at least amounts to some kind of afterlife. Maybe that afterlife is not necessarily pleasant. But whether we are then in heaven or hell (or lead a more shadowy "virtual" existence in a digital Hades as opposed to "real" physical life), we are no longer in this world. And this is precisely what our religious and apocalyptic minds desire. As Geraci has put it, "Apocalyptic AI promises its faithful a life of eternal reward in a virtual afterlife."[11] He gives the example of *Second Life*, a virtual world in which some people want to live forever. We may also consider other virtual worlds, game worlds, and technological visions such as the metaverse.[12]

Second, although we still have to return to our earthly existence when we use these worlds and games, some transhumanists such as Kurzweil believe that we will be able to upload our minds to computers and become "digitally immortal."[13] The idea is that someone's brain is scanned in order to emulate the mental state of the individual in a digital computer. The computer then runs a simulation of the brain's information processing. In this way, it is thought, the person's mind can be reinstantiated and digitally stored forever. Kurzweil thinks that the process would capture a person's personality, memory, skills, and history.[14] It is believed that people's minds and identities will be transferred into computers; it is assumed that they would still end up with sentient, conscious minds. The body is then obsolete; if needed at all, it can be replaced with mechanical parts, Kurzweil suggests. Biological brains will be converted into machine minds. He has predicted that by 2045 we will be able to upload ourselves.[15] Thus the idea is that we will technologically achieve perhaps not immortality of the soul but at least immortality of the mind. Or to put it in Gnostic language, the imperfect, corrupt, material, and mortal body can finally be left behind. The soul can leave its prison.

Third, it is already possible to create so-called deadbots or griefbots, which seems to amount to a "digital resurrection" of the dead. Deceased people are given an "AI afterlife" by having AI chatbots recreate their language patterns and personalities on the basis of their digital data (e.g., text messages, emails, social media, answers to personality questions, voice recordings, and so on).[16] Our data live on in digital space anyway, so why not use them? The main idea is to comfort mourners: by means of these "AI resurrections," people are given the illusion that it is now possible to talk to their loved ones who have passed away.[17] Some companies also allow people to see an AI-generated video that simulates the dead person. The family can then continue to interact with the digital version of the deceased, who may appear in the form of an avatar or a hologram. And maybe such deadbots or "AI ghosts" also help those who are going to die?[18] As Laurence Devillers puts it, "People have always wanted to be invincible, immortal. It's part of our founding myths—nobody wants to die."[19] A virtual avatar is then seen as, perhaps, a partial solution.

Fourth, beyond resurrection of the mind, the immortality of the body is also now on the agenda. Technologies are already used to increase lifespan, and now the (again transhumanist) idea is that they can also be used to radically extend life and, after death, to preserve and potentially resurrect bodies.

Transhumanists propose radical life extension and believe that technology will eventually eliminate aging or disease as causes of death. Aubrey de Grey and Michael Rae have argued that aging is a disease that can be reversed through medical interventions. Like other transhumanists, they believe that the limits of the human body must be transcended, including limits imposed by aging and death. Once we understand the biological processes that lead to aging, we can reverse them. We can repair damage to cells. This would enable people to live hundreds or thousands of years in good health.[20] AI could help to discover drugs and nutrition that target the biological mechanisms of aging. Kurzweil also counts on nanorobots that could treat diseases.[21] Few will want a longer life if that means prolonged suffering. But if we could have a long and healthy life, why not? This kind of longevity does not give us immortality but something that at least seems highly desirable: living a long(er) life.

Yet some transhumanists believe that even after death something can be done, that is to say, resurrection of the body by means of technology. Cryonics is the practice of preserving the bodies of people who have died from currently incurable illnesses at very low temperatures (e.g., –196°C). After death, the bodies are quickly cooled down. This slows down processes that lead to further cellular damage. A solution, a cryoprotectant, is used to prevent ice formation. AI could be used to improve cryopreservation methods. Nanobots may also help. The hope is that in the future new medical technology will enable these people to be revived and treated. As Cave has argued, cryonics can be viewed as a technological approach to achieving the afterlife: it is a version of resurrection.[22] But this time scientists of the future, not god(s), are supposed to resurrect people. Thiel already signed up. His view of death? "I'm against it."[23]

Like the tech kings, Gilgamesh too was "against death." Existentially speaking, we are all against death and most of us struggle against it in some way or other. It is hard to accept our own death and that of our loved ones. Perhaps it is indeed natural to fight death, as Thiel thinks.[24] We are, in a way, vulnerability fighters who use technologies as weapons in this fight, as I have suggested in my book *Human Being @ Risk*.[25] With our machines, we struggle against human vulnerability and mortality. AI is not only the new technology but also the new anti-vulnerability tool and anti-mortality cannon. But can technology save us? Can AI save us? Can a deus ex machina give us eternal life?

Transhumanists are optimistic. But there are many reasons for viewing these ideas and the technological practices that are meant to give us immortality as philosophically problematic, ethically controversial, scientifically speculative, and ultimately futile and failing in the face of human mortality. And these discussions touch on fundamental philosophical and religious questions regarding the nature of humans and their minds and, of course, the perennial existential and religious issue of (im)mortality.

*

Virtual worlds and computer games do not really enable us to escape our vulnerable and embodied condition. The problem is not just that current systems fail to provide the full range of sensory experience and physical interactions we are used to from the ordinary physical world, and thus in this sense fail to properly simulate living in a different world. The problem is more fundamental and has to do with assumptions about the relation between mind and body that underpin the idea that we can escape with our minds to another world.

Immersion in a virtual world, at least if understood as an escape from the body, seems to assume that we can separate body and mind, that the mind can be liberated from the body and roam freely in the virtual world while the body stays behind. But mind depends on body. Many cognitive scientists and psychologists today endorse the concept of embodied cognition: our cognitive processes, including consciousness and thinking, are intertwined with, and influenced by, our physical body. When we play games or interact in virtual worlds, however immersed we are in terms of experience, we are as much embodied there as anywhere else. And "in" virtual worlds we are as vulnerable and mortal as we are anywhere else. The promised escape, while available to everyone, is an illusion.

The idea of uploading equally assumes this (neo-)Cartesian dualist and reductionist notion that mind and body can somehow be separated and that mind does not need body.[26] It assumes that we can detach the mind and its conscious processes from the biological body while preserving the person and their consciousness. But reducing consciousness to data processing overlooks the emergent nature of the human mind, which is linked to the body and depends on physical processes. It is fundamentally irreducible to computational terms. Transhumanists tend to assume that minds are brains and that brains are computers. In philosophy of AI, these views have been famously questioned by Hubert Dreyfus, a philosopher who critically engaged with early AI research. Based on existential phenomenology (Martin Heidegger, Maurice Merleau-Ponty), Dreyfus has argued that computational models of the mind fail to account for the embodied, context-dependent, and nonformal aspects of human experience. Human beings understand their world through their bodily experiences and interactions with their environment, thus building up a background of implicit knowledge that cannot be reduced to rules or algorithms.[27]

Another philosopher, John Searle, conceived of the Chinese Room thought experiment to argue that computer programs may produce the *appearance* of understanding language but lack real understanding. In the thought experiment, Searle sits in a closed room and receives Chinese characters slipped under the door, to which he responds by using a computer program with rules for manipulating these symbols. Searle does not understand Chinese, but after using the rules and producing a valid output, the people outside mistakenly suppose there is a Chinese speaker in the room.[28] While today's AI is not rule based in the sense of the AI Dreyfus and Searle envisioned and discussed, instead using machine learning, programs such as ChatGPT are good at predicting text but do not really *understand* language or think, and neither do other computer programs. Uploading enthusiasts still seem to assume a computational model of the mind and fail to recognize the limits of what computers can do. And there seems to be no good reason to think that data processing and computer simulation could lead to consciousness and understanding as we know it in humans; only a reductionist understanding of consciousness and persons could support the very idea. Not to mention that religious people will not find a soul in the machine, let alone the soul of their loved ones.

Philosophically, it is indeed questionable whether identity would be preserved after uploading. It is doubtful that the uploaded and reinstated consciousness—if this were to be possible at all—would truly represent the original self given that it is a computer *simulation*. Would it mean that the original person still existed or that a new entity was created?[29] The latter seems more probable, at least if we think of personal identity in terms of continuity of consciousness, experience, and the physical body, a continuity that is unlikely to be preserved with uploading. It is equally questionable whether the subjective nature of human experience can be replicated at all. Would an uploaded consciousness feel the same as the original one, and should uploading be done if there is even the slightest doubt about that? How ethically acceptable would that be? And what would it be *like* to "be" a simulated mind, if this rather reductionist question makes sense at all given our embodied and holistic nature? Moreover, even if these fundamental hurdles could be overcome, the uploading idea is highly speculative; we do not have the science to do this, and there is no indication we will have it in the near future. Scientifically and philosophically speaking, our understanding of mind and consciousness is still very limited. Scientifically and technologically, we are nowhere close to capturing the complexity of the human brain or even to simulating the human brain. It is therefore highly unlikely that this technological immortality trick would work. It is also likely to be prohibitively expensive. Uploading may be only for the select few.

Note that most Christian theologians also reject dualism. While a minority see the body as entirely separate from the soul, the body is usually seen as an essential part of the person. The doctrine of incarnation is usually interpreted as implying that people are integrated wholes and that therefore resurrection implies that the whole person is raised and gets a new life in a transformed *body*. This is one reason why not only conservative but also more liberal theologians reject transhumanism, or at least ideas such as mind uploading. Christianity has also always rejected Gnostic dualism. And the Jewish religion affirms the psychosomatic unity of the person, a view that influenced Paul when he laid the foundation for core Christian beliefs.[30]

The objections against dualism and reductionism also entail that the arguably more democratic and cheaper creation of digital versions of dead people in the form of deadbots is doomed to failure, at least if it is meant to bring back the deceased. Consciousness cannot be reduced to digital terms, even in the form of conversational patterns. Perhaps the large language models that power the AI deadbots get us closer than ever to simulating a conversation with another human being. The illusion is being perfected; both LLMs and technologies that create digital simulations of persons are getting better. But as the bots, avatars, and holograms are based purely on digital traces, we will always miss the holistic, embodied person and the conversations with their embodied mind. Even if we are confronted with a spooky moving image of the deceased, the digital ghost is not the person and, in fact, is not even a real ghost in the traditional sense. If dualism were correct, ghosts might exist, since then it would be possible for mind to separate from body, soul from corpse, spirit from matter. Some of the religious views mentioned in this chapter might be interpreted as entailing such a dualism, even if they are often more complex. The creation of digital ghosts might then make more sense, since at least *in principle* we could separate mind from body, soul from physical substrate—of course leaving aside the technological difficulties. Perhaps then we would not need a griefbot in the first place because we could more directly connect with our diseased loved ones in the form of ghosts. But in spite of our religious hopes, dualism does not accurately describe reality. At best, digital resurrection provides temporary comfort. At worst, it is ethically highly problematic: there might be lack of consent to the use of the relevant personal data, or AI might not succeed in accurately representing the deceased individual, or the grieving process might be prolonged or complicated. People might have unrealistic expectations and get distressed or even traumatized by the experience of the image, voice, and interaction. It may at the least be very uncanny. In addition, digital personalities might be used to generate profit and may be exploited in ways that do not respect the dignity of the deceased. It is also unclear what the long-term impact of this technology might be on how societies deal with death. It may certainly contribute to Western efforts to try to deny and fight human mortality. But these efforts are bound to fail and may cause more suffering. Moreover, they may disrupt some potentially wise, traditional ways of dealing with the deaths of loved ones such as mourning rituals (and indeed wise ways of coping with the prospect of one's own death).

The dualism and reductionism objections also do not help the case of cryonics, which is often seen as a pseudoscience by nonbelievers. Most scientists regard it as highly improbable that corpses, even if preserved in this way, could be revived at all. There is no evidence for revival, cryoprotectants can be very toxic, and it is highly speculative that we could soon have technologies to repair the damage to cells. And whatever scientific reasons may be given, there is a good philosophical objection similar to that against uploading, which is again about dualism and reductionism: while mind is connected to body, we are not our physical bodies, or at least not *just* our physical bodies, and we are certainly not reducible to a *dead* body. Just as mind cannot be separated from body (uploading), body cannot be separated from mind (cryonics). Once the body dies, the mind dies as well. What, then, would be resurrected? Even the Christian idea of resurrection is usually not interpreted as the mere resuscitation of a dead body, instead entailing a "transformative" type of resurrection in which the spiritual body continues in a new, different way.[31] And as with uploading, it is also not clear if personal identity would be preserved with cryonics. Instead it seems, again, that a new entity would be created, not necessarily a Frankensteinian monster, perhaps, but in any case not the diseased person (which would be uncanny enough, given that there would be similarities in appearance). It is also not clear how society would deal with resurrected persons and vice versa. To wait for resurrection of cryonic corpses, then, seems rather like awaiting the resurrection of the ancient Egyptians: a bizarre hobby, for sure, and certainly both costly and futile. But, of course, a project that is very understandable in light of the history of religious ideas and indeed the history of humanity. We keep struggling. We keep fighting. We wait. We hope. We don't give up. And like with the ancient Egyptians, it's only the rich and powerful that might be preserved. The rest of us are supposed to rot.

Finally, the very idea of the deus ex machina, and more generally of a god that is supposed to save us, can be helpfully criticized by using Nietzsche and by appealing to human responsibility: it is a dangerous illusion to hope for a god to save us, to deny our human freedom, and to evade responsibility for our lives, relationships, and societies on earth.[32] If a particular religion embodies and nurtures that hope, one could argue that we don't need that religion (or that interpretation of the religion in question)—nor do we need the "secular religion" of transhumanism that deifies technology. Transhumanists, the majority of whom tend to see themselves as secular scientists and philosophers, are more steeped in Christianity than they realize or are willing to acknowledge. And to the extent that we all hope for technologies to save us from mortality, we are too. Nietzsche criticized Christianity for promoting passivity and dependence on a divine savior. Christians, he argued, have a slave morality and wait for divine salvation and redemption rather than creating their own destiny.[33] They also have an otherworldly focus, which leads to a rejection of the world as it is. Instead of waiting for a god to help or hoping for an afterlife, digital or not, Nietzsche's view suggests that we should live in the here and now, take our lives in our own hands, and make the best of them—transcending ourselves and perhaps working on our legacy (whatever that may mean), but *within* the limitations of the human condition.[34] Gilgamesh got that, in the end.

8
Conclusion: Deus Ex Machina, or Why We (Think We) Need Machines

Saying that AI is "artificial religion" is mainly meant metaphorically. AI technologies and their related communities are not literally religions. But the metaphorical use of the term works and is legitimate. Consider the term *AI* itself: "artificial intelligence" is also a metaphor in that our machines are not really intelligent, at least not in the way that humans are. Yet the term makes sense because the idea since the 1956 Dartmouth College workshop that launched the field has been to simulate human intelligence.[1] The term *artificial religion* also makes sense. In this book, I have argued that AI—as an intellectual project, public discourse, and technological practice—is rooted in, and still influenced by, Western religious culture. Even in so-called secular societies and cultures, it is important to acknowledge and study these roots in order to better understand how we think and talk about our intelligent machines and how we interact with them, and to gain a critical relation to these ways of thinking and doing. Throughout this book, I have shown that patterns of religious and mythological thinking and practice continue today and shape AI and our relation to it. Religious narratives about automata for the gods, abandoned creatures, wondrous machines, gods we can talk to, apocalypses, and immortality *pharmaka* are still relevant to today's existential struggles, and they also shed some light on the ways we speak about and interact with AI and other contemporary "intelligent" machines. Even today's supposedly secular policy discourse concerning AI is still influenced by religious thinking understood in this broad sense, for example when people buy into transhumanist thinking and assume apocalyptic AI scenarios.

I emphasize this "broad sense" of religious thinking since these myths and ideas are not *only* religious in the narrow sense of being connected to the big, institutionalized religions of the past and present—here especially the ancient Greek, Jewish, and Christian religions. The narratives and ideas that figure in this book were developed for and found their specific cultural expressions in these specific religious and historical contexts, and in turn have their specific enduring influence and concrete impact in particular cultural and political contexts such as Silicon Valley. These specific expressions are perhaps typically Western. But they all touch on some more general, and arguably deeper, existential and spiritual concerns, struggles, and aspirations that belong to us as *human beings* whether or not we identify as religious or are members of a particular religion. The narratives we have encountered are not just about gods and divine matters. They are also about people: people dreaming of not having to work, people looking for security in an insecure world, people struggling with how to relate to their offspring (and vice versa), people wondering at the world (including the world of science and technology), people longing for a relational other and craving companionship, friendship, and love, people desiring to know their personal future and the future of the world, and people who do not want to die and who hope for at least some form of immortality. These are not just religious problems and aspirations (if understood in a narrow sense) but also belong to human existence as such. They are universal human problems and aspirations.

Interestingly and perhaps surprisingly, it is both these religious and mythological ways of thinking *and* these underlying existential struggles and hopes that begin to give answers to my initial question about how to understand the puzzling Western relation to machines, and, in particular, to questions that have been on my mind since I started working on the ethics of robotics and AI many years ago: Why do we build and use intelligent machines at all? Why do we (think we) need machines? Where does that "need" come from? Where does that dream come from?

The answers can be found in the aforementioned religious narratives and patterns of thinking that have shaped Western machine imagination, including Western AI imagination, in specific and unique ways. To put it crudely, to really understand Silicon Valley, you have to understand Jerusalem, Athens, and Rome. To *really* understand and contextualize what people like Musk, Zuckerberg, and Altman are doing and why they find many followers and gain power, you have to understand Hephaestus, Rabbi Loew, and Jesus Christ. Contemporary technologies such as AI and the aspirations and imaginaries of their developers and users can only be fully understood in the light of a broader intellectual and cultural history. Understanding Western technologies requires understanding Western culture, including Western myths and religions. In this book, I have lifted only a corner of that tapestry, revealing just a glimpse of the intricate patterns beneath. As my endnotes suggest, often more detailed scholarly work has already been done on each theme. And more work needs to be done: the full masterpiece of this fascinating religious-technological history remains to be unveiled. In the previous chapters I have offered a selection of some of the most relevant narratives and imaginaries that introduce this history and help us interpret our relation to AI and similar technologies.

But the answers to the question of why we think we need intelligent machines can also be found in the deeper existential needs and struggles that arguably inspired our religious narratives and to which the related religious practices, thinking, writings, and institutions are a response, whatever else they may be. At least one, and perhaps the most fundamental, set of explanations for why we build, use, and dream of intelligent machines such as AI is that we: don't want to work; seek protection; want agency and power; want children or some kind of offspring at least (artificial if needed); crave wonder and mystery; seek communication, companionship, love, and guidance; desire to know the future; look for anything or anyone that offers hope for the future; and want to live forever with our loved ones. Again, these are not just "religious" concerns and aspirations. They are linked to deeper existential structures and relations that belong to being human. They may partly constitute matters of "ultimate concern," to return to this influential expression from Tillich, and they certainly are themes that religions have always grappled with.[2] Consider again the problems regarding creatures and creators or the question regarding immortality, for instance. And these concerns and aspirations take on a specific cultural form, a Western form, for instance. But they are also at the same time desires, needs, and aspirations that mark human existence as such.

Technologies such as AI are supposed to be the answer to these all-too-human desires, needs, and aspirations. The Western ambitions, hopes, and dreams related to AI are in this sense of existential significance and origin. AI is meant to take over work we don't want to do. AI is embraced in military contexts to automate warfare and enhance soldiers. AI is meant to empower individuals and boost the power of collectives. AI is seen as a step toward "mind children" that may rebel but also continue our journey. AI chatbots are proposed as our new companions, friends, or even lovers, thus supposedly answering our existential yearning for communication with a relational other. AI is perceived as the new oracle that will guide us and predict the future. Moreover, AI is said to significantly shape that future: to open up a new future when everything will be different. And we hope that in the future AI may enable us to live much longer, preferably forever. As a technology but also as a narrative, discourse, myth, and imaginary, AI thus embodies many of the most fundamental existential aspirations, struggles, fears, and hopes of humankind. This renders the technology not only technically and functionally powerful (as an all-purpose technology) but also culturally powerful. It explains why AI has come to be seen as *the* defining technology of our time and times to come. It enables AI to appear as the technology that *shapes* our time and future.

Yet in contrast to the prevailing deterministic narratives about it that haunt Silicon Valley, AI—as technology and as cultural phenomenon—is not an alien "something" that does its work and has influence without us humans. Next to the "where" and "why" questions regarding AI's roots in religious thinking, ancient myths, and existential issues, we also need to ask "who" questions. The questions "Where does that (perceived) need for machines come from?" and "Why do we dream of intelligent machines?" also have such a "who" answer. AI is always linked to humans and to concrete political and socioeconomic contexts, interests, and agendas.

There are people who *want* us to need and buy these intelligent machines and who have them designed for that purpose. There are people who use the machines, get used to them, and then get used *by* them—and perhaps also by those who create and own the technology. There are people who want us to see AI as a deus ex machina and a solution to all our existential challenges and religious aspirations since that makes AI sell. There are people who link AI to mythologies that hide their own role and responsibility, presenting AI as a thing out there and as an unavoidable destiny. Asking why humans develop and use AI is thus not only a question that requires an interpretive, hermeneutic exercise but also a question that leads us to more questions regarding politics and power. Throughout the chapters of this book, I have shown that the relevant religious narratives, patterns of thinking, and existential issues also have a political dimension.

This is not only so because, as I have sometimes noted in this book, religions and religious thinking have a political dimension. Consider, for instance, slavery thinking in ancient Greek and Egyptian times, or Christian disciplinary practices and surveillance techniques. These ways of thinking are also often continued in contemporary thought and talk about AI, for example when we want AI to be our slave or when modern nation-states use AI to discipline us. Politics is part of what AI is and does today, and it was always a dimension of religion. The current public discourse about AI has a political-religious dimension, and our AI practices and hopes concerning AI are shaped by ancient myths and patterns of thinking and practices that have been developed in specific political-religious contexts. Today Big Tech creates new practices of confession, surveillance, and disciplining, and is building new pyramids of power and resurrection for the elite. If we know our political-religious history and its related ideas, we can see this power dimension and have appropriate critical distance and reactions.

Politics is also an unavoidable aspect of how the existential issues play out in our actual social and political reality, thus connecting to political economies and questions concerning justice. We think we need machines, or are made to believe that we need them, for all these religious and existential reasons. But not everyone gets what they want, and different people may get different things in different political and economic contexts. Different people, groups, and classes are impacted in different ways by AI. "We" might not benefit from AI, or at least not all of us in the same way. Not everyone gets their robot slave or is equally empowered by AI. Some people—owners of capital, members of the techno-elite—become incredibly wealthy or influential; others remain human workers (data serfs?) or lose their jobs and thereby their hope for a basic and decent level of material and social security. What is *their* future? AI is widely seen as having the potential to exacerbate existing inequalities, widening the gap between rich and poor. Who will win and who will lose in the hegemonic, colonial, and eschatological AI scenarios? What if AI leads to worse social and economic conditions for the majority? What if Earth becomes increasingly unlivable? Technological enhancement or escape to space may be available to only the rich and powerful. What about those who are not enhanced and cannot join the cosmic, superintelligence adventures and colonial expeditions of billionaires but have to stay on Earth? Will these people provide the raw materials, do the dirty work, or populate a human zoo for entertaining the enhanced? (If this sounds far-fetched, consider that this is what we currently do to many less-intelligent animals. Or consider the dystopian series *The Hunger Games* and *Squid Game*. This divide is already part of the Western imagination and, unfortunately, Western practices.) Are ordinary mortals still needed and useful at all? Are they supposed to go extinct? And what about those who already *today* are not admitted to AI wonderland but have to labor under bad conditions to create the data, devices, and energy that fuel the AI paradise for the few? Who will benefit from the Brave New AI World? Who will be the kings and who the servants or slaves? Who is and will be in control of the artificial offspring? Who is entertained and tricked by AI, and who profits from that? Who is excluded from the magic show? Who is distracted by it? Who is exploited by chatbots and who benefits from that exploitation? Who is and will be manipulated, controlled, and disciplined on the basis of AI predictions? Is AI supporting new forms of capitalism or should we talk about feudalism again, as some propose? Have the owners of Big Tech become our new feudal overlords?[3] Is AI mainly a Western project or at least reserved for the big geopolitical and technopolitical powers of this world, or will it also benefit the Global South? Which prophets, companies, and politicians benefit from mongering in doom narratives and eschatological hopes about AI, and what are their agendas, interests, and influences on policymaking and tech investment? Who present themselves as saviors, protectors, and guides of nations and humankind? Who will survive the AI apocalypse and who will be doomed to perish? These questions regarding the future of and with AI are not only about the future of humankind but also about the present and future of actual humans and groups of humans within specific political-economic conditions, structures, and contexts—which in turn may be significantly transformed by AI.

What is at stake, then, in discussions about AI in general but also when we ask why "we" develop and use AI and other intelligent machines in the first place is not simply a specific technology and its future. It is also our *political* future. Behind the deus ex machina that is presented to us as a universal solution for our problems there are people, and these people have agendas, interests, and power. Understanding why we develop and use AI and developing a *critical* relation to the discourse and imaginaries about AI requires not only a deep dive into the history of religious ideas and practices—and a study of their continuation in today's thinking about the technology and today's technological practices—but also a more systematic study of the relations between AI, religion, and *power*. In this book I have only provided some insights and suggestions regarding this issue; here, too, more work is needed. We also need more political awareness about these matters. If we do not pay enough attention, we are likely to sleepwalk into a future where, through the use of AI, our deepest religious and existential concerns are exploited to an extent that history has never seen before.

Know Thyself

The Socratic and originally Delphic imperative "know thyself" is an ancient spiritual aphorism that is also a cornerstone of Western philosophy. In the context of thinking about artificial intelligence, it means at least two things, which concern two kinds of ignorance: ignorance about our cultural and religious roots and ignorance about politics.

First, in order to really know what we are saying about AI and what we are doing with AI, we have to know what AI is and what "we" are. Indeed, knowing AI entails not only knowing the technology as such, but also knowing the cultural and religious roots of what we say and do. We have to know the stories, the myths. We have to know how our ancestors thought. And we have to know how those ways of thinking have not only found expression in specific cultural and historical contexts but also ultimately responded to key existential and universal concerns we have as human beings. In this sense, knowing AI requires us to get to know *ourselves*. Thus the question concerning AI is not only about technology but also about us, about humans, about human existence, and about concrete cultural and religious traditions that relate to that human existence in specific ways. Once we realize this, we can embark on a journey that may give us new insights into our relation to AI and, more generally, into our contemporary technological condition. Acknowledging human finitude and trying to develop technologies that help us cope with this finitude rather than try to overcome it, for example, could be one outcome of this journey. Or learning from a specific tradition how to deal with creational relations in nontoxic ways could be another outcome. In stories we may find inspiration for an ethics of technology, for example. We might want to apply that ethics to the use and development of AI. But from Socrates and his spiritual source, the oracle at Delphi, we can learn that self-knowledge and knowledge of one's own limitations precedes ethics and is, rather, a precondition for ethics. We have to start with acknowledging our ignorance about cultural and religious matters in order to be fully open to learning from the history of ideas and to become aware that these ideas may (still) be shaping our contemporary thinking and our contemporary ways of relating to our existence, including ways of developing, using, and talking about technology. Many people who talk about AI are ignorant about the cultural and religious background influences and structures that shape their thinking. (Elsewhere I have called these background structures AI's "grammar," and earlier in this book I have also referred to them as the "collective 'religious unconsciousness'" and as patterns of thinking that still "haunt" us.[4]) In this sense, they do not know what they are saying or doing. A first step is to recognize this ignorance; then a sufficiently critical ethics of technology—and ultimately the development of truly *ethical AI* and other ethical technologies—is possible.

When it comes to acknowledging the religious background of AI, let me add a note on what we may further conclude from this with regard to the discussion about secularization. Since this book reveals AI's religious grammar, some may interpret it as a contribution to the project of the demythologization and secularization of AI in the sense that they may want to strip AI of these patterns and influences. I sympathize with this project. This book can be read as a criticism of AI religion, inspired by Feuerbach and Nietzsche, among others. But I also have to issue a warning. Whether or not demythologization and secularization is desirable, it is doubtful to what extent it is possible. While we can and should certainly take a lot more critical distance, it would be an illusion to think that we can entirely purify AI of religion, myth, and other cultural influences.

Myth and religion are here to stay and continue to exert their influence, on AI as elsewhere. If Big Tech manages to rely on these narratives, patterns, and influences in order to maintain its power and manipulate us, it is because they are already present in the public, and they are in the public because they are deeply entrenched in Western cultural grammar and carefully and intricately connected with our deepest existential needs. The challenge, therefore, is in the first instance less about achieving total purification of these narratives and more about *which* narratives we want (or not) and how we can relate more critically to our own collective religious patterns of thinking.

Paradoxically, a true secularization of technology (if such a thing were possible and desirable) requires more, not less, intellectual engagement with those patterns of thinking. Those who seek to establish a secular "tech humanism" or a secular "digital humanism," for instance, need to study the cultural-religious background of tech in order to better formulate their own project.[5] It also seems to me that we will always need myths to make sense of the world and to make meaning, even and also in the modern world—a thought that is in line with Nietzsche, Jung, Heidegger, Blumenberg, Campbell, and Ricoeur, among others. The fact that some of these myths have been and are still used to oppress people, support war, and galvanize colonization is all the more reason to critically study them and to try to create different myths or offer different interpretations of these myths.

Myths, like technologies, are political. Avoiding the topic does not help to tackle the problems and explore the good possibilities that myths create. Moreover, attention to myth can also make us aware of the limits of AI, helping us to avoid the hubris of thinking that everything can be calculated, controlled, and predicted and, ultimately, that we can be (like) gods.[6] And if I'm right about the deep entanglement of technology and culture, then to imagine truly new technologies means to imagine a new culture—something that is always possible only to a limited extent given that we are steeped in our own. Modernity has erred insofar as it has cultivated ignorance about this limitation, falsely suggesting that it is possible to start from scratch. Ethics of AI, like all ethics of technology, therefore has to start from somewhere. And this shared "somewhere" requires collective self-examination.

That being said, this book partly demythologizes AI in the sense that by describing its underlying religious narratives and patterns of thinking, it mitigates ignorance about their continuing influence and enables us to gain distance from these patterns, and indeed from those who employ them to their advantage. For example, we may want to criticize patterns such as waiting for AI to save us, seeing AI as a destructive, malicious god, and more generally ascribing godlike qualities to AI. We may also want to criticize the slave-master dream, the apocalyptic superintelligence narrative, the illusion of artificial companionship, or the myth of technological immortality. We can get at least *some* distance from such narratives as long as we know that they are deeply rooted in Western cultural and religious traditions and that it is not easy to leave them behind, and as long as we recognize that they are likely to continue exercising at least some of their seductive power and deep, hidden influence on our technological culture. But there is no need to despair; change is possible to some extent. AI is not a god unless we make it one (unfortunately we seem to be prone to doing so). As I continue to emphasize in other work on AI, AI is always linked to humans and therefore there is no reason to be determinist or fatalist about it. Given that humans develop and use AI, we can in principle steer a different course and bend AI and other technologies in different, better directions. This premise renders an ethics of AI possible; rejecting technodeterminism is another precondition of AI ethics.

Second, however, bringing about this change has its own challenges and leads us to a second kind of ignorance, which is an ignorance about politics. We should also realize that we are often *politically* ignorant and naive—especially when it comes to the theme of technology, which at first sight seems unrelated to politics. In modernity we are used to categorically separating politics from nature, or separating thinking about humans, culture, and power from physics, engineering, and computer science. But this way of thinking is mistaken and even dangerous. Human existence has an unavoidable social and political dimension, as does our thinking about the use and development of technology (including ethical thinking about technology). Here the path of self-examination and self-knowledge is not about getting to know human beings and human existence in the abstract, or about studying a particular philosophical or religious tradition and its myths as such, but about examining and acknowledging the political aspects of these ways of thinking and, more fundamentally, achieving awareness about how we as social beings are always situated in a particular context of power relations.[7] When we deal with human existence in some way or another, and when we create technology and narratives about technology, we are related to others and are part of a particular political, social, and economic context, which always includes its own power dynamics. If we wish to use or develop or talk about AI, then we have to recognize that we are doing so from such a particular political context, for example one shaped by capitalism and neoliberalism and one that is situated in a particular part of the world (e.g., the West, the North, etc.). Whatever we do with or say about AI may be influenced by this context and will always somehow interact with a field of interests and power.

If we are ignorant about this political dimension of AI, then we can be easily manipulated or controlled by others, exploited by the systems we live in (and by those who benefit from those systems), or led to do or say things that may seem right or true in our own universe of thinking but that might not be sufficiently responsive to those we address with our discourse or create technology for. When we talk about or develop AI, others may talk back. Communication, like technology, is always politically risky. Even academic texts cannot avoid this risk. Many people who develop AI or own AI companies are politically ignorant, which gets them into trouble when what they say or develop lands in a political minefield outside their own bubble. Here, knowing AI, understood as knowing yourself, means knowing your own political and economic situation in relation to others and the socioeconomic system (e.g., your situation as an AI developer or AI user, and your situation as worker, manager, or owner within AI capitalism). For example, many participants in the public discussion about AI speak about it from a privileged position. Heeding Marx, one could add that knowing yourself means knowing your class. The Industrial Revolution was not only about technologies but also about ideas and social relations. We should be aware that even today, and via AI, the persistence of ancient religious patterns of thinking and universally human existential weaknesses can be used to legitimize and maintain a particular social order (e.g., capitalism, colonialism, patriarchy, etc.), perpetuate specific socioeconomic relations, and exploit and exclude people. It can also be used to support harmful ideologies and problematic worldviews. Any kind of humanism that wants to do something about this needs to gain sufficient self-knowledge about the relevant socioeconomic and cultural background and its entanglements with both technology and power—including some of the problematic background and entanglements of humanism itself, such as strong forms of anthropocentrism. Indeed, some of these patterns of thinking and technologies have led, and still lead, to exploitation of and violence toward nonhumans, environmental degradation, and climate change. A humanism for the twenty-first century can only be a humanism that questions any strong form of anthropocentrism, as I argue elsewhere.[8] We have to reflect on who we are, where we stand, and where we are heading with regard to those issues. The digital transformation is entangled with social, political, and environmental changes; we (as thinkers about AI) and the discourse about AI are situated in the midst of these storms and transformations.

Without self-awareness and self-knowledge in these two senses—cultural and political—the use and development of AI and the discourse about AI will still be connected to and shaped by these cultural patterns and power structures but lack critical distance and reflection. A lack of education on these matters may lead to what we, paraphrasing Nietzsche, call today a "herd mentality."[9] If we use AI-powered social media to voice our opinion about AI and participate in the public AI debate but lack both cultural background and political awareness, we risk simply parroting whatever others say and do. When posting about AI or when making claims about the future of AI, we then merely follow the herd. We become mere avatars of lifeless thoughts, fossilized values, and vested interests. In other words, we come to resemble our AI chatbots. Then, indeed, as herd animals without any critical resources, we have put ourselves in a situation in which only a machine, or god, can save us.

The German philosopher Hans-Georg Gadamer, a key figure in twentieth-century hermeneutics, might have been right that hope is a fundamental aspect of human existence.[10] It certainly seems to be a key feature of Western culture. But heeding Nietzsche and Marx we should also be aware that hope can be highly problematic. Hope may not lead to change and might even be a barrier to it: it may be an opium that supports the status quo. The specific form that hope takes in Western religious and philosophical traditions and, influenced by those traditions, in contemporary AI discourse, risks becoming a narcotic. It might *prevent* true liberation and transcendence—not via transhumanism or a Singularity, but the kind of transcendence Nietzsche thought was possible for us mortals.[11] And it could lead us slowly but surely into a condition of a universal herd mentality, slavery, and domination, enabled by AI and other advanced technologies misleadingly promising salvation and guided by the false prophets of the tech apocalypse.

We should take action to avoid such a dire situation in which we can only hope for a deus ex machina to save us (and perhaps, if that goes wrong, for another deus ex machina to save us from the first). More intellectual work on understanding the deeper roots of our technological cultures (Western and others) and discussions about what they mean for AI, along with education that increases cultural knowledge and raises political awareness with regard to contemporary technologies and society, can contribute to a more critical relation to AI. More generally, such intellectual work and education may help humankind achieve the independent thinking that Kant thought was needed for enlightenment and avoid the herdlike mentality that Nietzsche (and later many twentieth-century thinkers) criticized so much. In order to know ourselves, today, we have to dare to know (*sapere aude*) who and what we are and have become as the result of religious narratives, political history, and modern technologies. Paradoxically, perhaps, recognizing the *limits* of our autonomy in this sense may give us the critical distance and freedom not only to develop new, more independent thinking but also to imagine the technologies we really need.

Notes

Chapter 1

1. For more discussion of these questions, see Mark Coeckelbergh and David Gunkel, *Communicative AI: A Critical Introduction to Large Language Models* (Polity Press, 2025).

2. Mario Klingemann, *Appropriate Response*, 2020, https://www.artsy.net/artwork/mario-klingemann-appropriate-response-1. See also Marcelo Soria-Rodríguez, "On Mario Klingemann's Appropriate Response," *i-illucid* (blog), February 21, 2021, https://www.iillucid.com/on-mario-klingemanns-appropriate-response/.

3. Yuval Harari, *Homo Deus* (Vintage Penguin Random House, 2015); Vatican, "Rome Call for AI Ethics," February 28, 2020, https://www.vatican.va/roman_curia/pontifical_academies/acdlife/documents/rc_pont-acd_life_doc_20202228_rome-call-for-ai-ethics_en.pdf.

4. See for example the religious tropes touched on by Émile Torres in his criticism of longtermism and affective altruism in Andrew Anthony, "'What If Everybody Decided Not to Have Children': The Philosopher Questioning Humanity's Future," *The Guardian*, July 22, 2023, https://www.theguardian.com/books/2023/jul/22/pro-extinctionis-longtermim-effective-altruism-human-extinction-emile-torres.

5. Greg M. Epstein, *Tech Agnostic: How Technology Became the World's Most Powerful Religion, and Why It Desperately Needs a Reformation* (MIT Press, 2024). Epstein is humanist chaplain at Harvard and MIT.

6. See for example Gary Marcus, *Taming Silicon Valley* (MIT Press, 2024); Arvind Narayanan and Sayash Kapoor, *AI Snake Oil: What Artificial Intelligence Can Do, What It Can't, and How to Tell the Difference* (Princeton University Press, 2024). It is also interesting that the very term "artificial intelligence" was coined by AI pioneer John McCarthy to get funding. Later, "artificial general intelligence" was created as a marketing term. See for example Brian Merchant, "AI Generated Business: The Rise of AGI and the Rush to Find a Working Revenue Model," AI Now Institute, December 2024, https://ainowinstitute.org/general/ai-generated-business.

7. Hans Blumenberg, *Work on Myth* (MIT Press, 1985).

8. For a more elaborate interpretation and development of Blumenberg, see Mark Coeckelbergh, "Myth, *Angst,* and AI: Towards a Neo-Blumenbergian Framework for Understanding How We Think About Technology," *Postdigital Science and Education* (2025): https://doi.org/10.1007/s42438-025-00548-x.

9. Paul Tillich, *Dynamics of Faith* (Harper & Row, 1957).

10. On AI's use by world religions, see for example Beth Singler, *Religion and Artificial Intelligence: An Introduction* (Routledge, 2025). In order to further explore the question of whether AI is (becoming) a religion, one could for instance use Ninian Smart's famous dimensions of religion. According to Smart, it is difficult and undesirable to give an essentialist definition of religion. Instead, he argues that religions have ritual, experiential-emotional, narrative or mythic, doctrinal or philosophical, ethical and legal, social and institutional, and material dimensions. Ninian Smart, *The World's Religions* (Cambridge University Press, 1998), 11–21. These dimensions can then be used to evaluate whether and to what extent AI, or at least some movements related to AI, are a religion. Such a project would not be very distant from Smart's own approach, given that he also analyzed secular worldviews and movements such as nationalism and Marxism.

11. Compare with the title of Nietzsche's book, *The Birth of Tragedy Out of the Spirit of Music* (Penguin Classics, 1994).

12. Peter L. Berger, *The Desecularization of the World: Resurgent Religion and World Politics* (Ethics and Public Policy Center, 1999).

13. See Nietzsche's announcement in *The Gay Science* (trans. Walter Kaufmannn [Vintage Books, 1974]) that God is dead and his view that in the absence of religion we need to create our own new values in *Thus Spoke Zarathustra: A Book for Everyone and No One* (trans. R. J. Hollingdale [Penguin Classics, 2003]).

14. Max Weber, "Science as a Vocation," in *From Max Weber: Essays in Sociology*, ed. H. H. Gerth and C. Wright Mills (Oxford University Press, 1946), 129–156.

15. On science and wonder see for example Richard Holmes, *The Age of Wonder: How the Romantic Generation Discovered the Beauty and Terror of Science* (HarperPress, 2009).

16. Such a notion of "religious unconsciousness" is different from, but could be inspired by, Jung's notion of the "collective unconsciousness," which argues that there are universal psychological structures (archetypes) that shape religion, or by work in sociology and anthropology that shows how religious-like patterns also persist in secular contexts. Consider for example Émile Durkheim's view that collective rituals perform similar functions to religious rituals or Edward Bailey's work on "implicit religion." See Émile Durkheim, *The Elementary Forms of Religious Life*, trans. Karen E. Fields (The Free Press, 1995); and Edward I. Bailey, *Implicit Religion: An Introduction* (Middlesex University Press, 1998).

17. See also Mark Coeckelbergh, *New Romantic Cyborgs: Romanticism, Information Technology, and the End of the Machine* (MIT Press, 2017).

18. Neil Postman, *Technopoly: The Surrender of Culture to Technology* (Knopf, 1992).

19. Ian Bogost, "The Cathedral of Computation," *The Atlantic*, January 15, 2015, https://www.theatlantic.com/technology/archive/2015/01/the-cathedral-of-computation/384300/.

20. David W. Noble, *The Religion of Technology: The Divinity of Man and the Spirit of Invention* (A. A. Knopf/Random House, 1997).

21. Robert M. Geraci, "Spiritual Robots: Religion and Our Scientific View of the Natural World," *Theology and Science* 4, no. 3 (2007): 229–246.

22. Singler, *Religion and Artificial Intelligence*, 17, 138. Singler sees new religious movements as a counterexample to the secularization thesis.

23. Taylor has argued for respecting religious identities in pluralistic societies and has explored religious sources of flourishing and morality. Jürgen Habermas has recognized that religion continues to play a public and political role and argued that we should include religious voices in the public sphere where religious and secular worldviews can coexist. Charles Taylor, *A Secular Age* (Harvard University Press, 2007); Jürgen Habermas, "Religion in the Public Sphere," *European Journal of Philosophy* 14, no. 1 (2006): 1–25.

24. See for example Mark Coeckelbergh, "The Grammars of AI: Towards a Structuralist and Transcendental Hermeneutics of Digital Technologies," *Technology and Language* 3, no. 2 (2022): 148–161; "Technology Games: Using Wittgenstein for Understanding and Evaluating Technology," *Science and Engineering Ethics* 24, no. 5 (2018): 1503–1519. The relevant work by Wittgenstein is *Philosophical Investigations*, trans. G. E. M. Anscombe, P. M. S. Hacker, and Joachim Schulte (Blackwell, 2009).

25. Marshall McLuhan, *Understanding Media: The Extensions of Man* (McGraw-Hill, 1964).

26. Epstein explores such literal and metaphorical connections in his work on "tech religion" in *Tech Agnostic*.

27. An example of the latter approach in contemporary cognitive science of religion is Justine E. Lane, *Understanding Religion Through Artificial Intelligence* (Bloomsbury Academic, 2021). Lane is interested in how computational modeling can offer an understanding of how groups maintain social cohesion, for example through ritual. Early cyberneticians have also been interested in religion. In chapter 3, I will discuss Norbert Wiener's work.

28. See for example the first part of *The Cambridge Companion to Religion and Artificial Intelligence*, ed. Beth Singler and Fraser Watts (Cambridge University Press, 2024).

29. Hans Moravec, *Mind Children: The Future of Robot and Human Intelligence* (Harvard University Press, 1988).

30. For work on how religions already respond to AI, see again Singler, *Religion and Artificial Intelligence*. One could also do collaborative work on religious narratives, rituals, and behaviors in AI use, thus doing for AI what Heidi Campbell and coauthors have done for gaming. See Heidi A. Campbell, Rachel Wagner, Shanny Luft, Rabia Gregory, Gregory Price Grieve, and Xenia Zeiler, "Gaming Religionworlds: Why Religious Studies Should Pay Attention to Religion in Gaming," *Journal of the American Academy of Religion* 84, no. 3 (2016): 641–664. For those interested in how ethical questions regarding AI and other contemporary technologies can be approached from various religious angles (in this case Christianity, Islam, and Judaism), I recommend Noreen Herzfeld, *Technology and Religion: Remaining Human in a Co-Created World* (Templeton Press, 2009). Readers are also encouraged to follow up with other references I provide throughout the book.

Chapter 2

1. Adrienne Mayor, *Gods and Robots: Myths, Machines, and Ancient Dreams of Technology* (Princeton University Press, 2018), 7, 1.

2. Homer, *The Iliad*, trans. Robert Fagles (Penguin Books, 1991), bks. 18, 5, and 8; Homer, *The Odyssey*, trans. Robert Fagles (Penguin Books, 1996), bk. 7.

3. Homer, *The Iliad*, 479.

4. Artistotle, *De Anima*, trans. J. A. Smith, in *The Complete Works of Aristotle*, rev. Oxford translation, vol. 1, ed. Jonathan Barnes (Princeton University Press, 1984), 641–692, ll. 406b17–22.

5. Artistotle, *Politics*, trans. Benjamin Jowett, in *The Complete Works of Aristotle*, rev. Oxford translation, vol. 2, ed. Jonathan Barnes (Princeton University Press, 1984), 1986–2129, l. 1253b.

6. Homer, *The Iliad*, bk. 18, 481.

7. Hesiod, *Works and Days*, in *The Homeric Hymns and Homerica*, trans. Hugh G. Evelyn-White (Harvard University Press, 1914), ll. 59–81.

8. Homer, *Odyssey*, bk. 8.

9. Homer, *Odyssey*, 539.

10. Philip Hefner, "Technology and Human Becoming," *Zygon: Journal of Religion & Science* 37, no. 3 (2002): 655–665, 657.

11. Plato, *Meno,* trans. G. M. A. Grube, in *Plato, Five Dialogues,* 2nd ed., rev. by John M. Cooper (Hackett Publishing, 2002), 58–92, ll. 97e, 98a.

12. Aristotle, *Nicomachean Ethics*, in *The Complete Works of Aristotle*, ed. Jonathan Barnes (Princeton University Press, 1984), 1729–1867, ll. 1161b2–7.

13. Genevieve Liveley and Sam Thomas, "Homer's Intelligent Machines: AI in Antiquity," in *AI Narratives: A History of Imaginative Thinking About Intelligent Machines*, ed. Stephen Cave, Kanta Dihal, and Sarah Dillon (Oxford University Press, 2020), 41.

14. Aristotle, *Politics*, 1986–2129, ll. 1253b34–1254a1.

15. See for instance Kevin LaGrandeur, "The Persistent Peril of the Artificial Slave," *Science Fiction Studies* 38, no. 2 (2011): 232–252.

16. Georg W. F. Hegel, *Phenomenology of the Spirit*, trans. A. V. Miller (Oxford University Press, 1977); Mark Coeckelbergh, "The Tragedy of the Master: Automation, Vulnerability, and Distance," *Ethics and Information Technology* 17 (2015): 219–229.

17. Herbert Marcuse, *One-Dimensional Man* (Beacon Press, 1964).

18. Shoshana Zuboff, *The Age of Surveillance Capitalism: The Fight for a Human Future at the New Frontier of Power* (Public Affairs, 2018).

19. Thomas Hobbes, *Leviathan* (Penguin Books, 1968), 168–169.

20. David Hume, *Dialogues Concerning Natural Religion and the Natural History of Religion*, ed. J. C. A. Gaskin (Oxford University Press, 1983), 26–32.

21. Karl Marx, introduction to "A Contribution to the Critique of Hegel's *Philosophy of Right*," in *Critique of Hegel's Philosophy of Right*, ed. Joseph O'Malley, trans. Annette Jolin and Joseph O'Malley (Oxford University Press, 1970).

22. Ludwig Feuerbach, *The Essence of Christianity*, trans. George Eliot (Harper & Brothers Publishers, 1957).

23. Max Weber, *The Protestant Ethic and the Spirit of Capitalism*, trans. Peter Baehr and Gordon C. Wells (Penguin, 2002).

24. Friedrich Nietzsche, *On the Genealogy of Morals*, trans. Douglas Smith (Oxford University Press, 1996), 114.

25. Friedrich Nietzsche, *Beyond Good and Evil*, trans. R. J. Hollingdale (Penguin, 2003), §46. He also argues in §202 that democrats want the autonomous herd.

26. Friedrich Nietzsche, "Expeditions of an Untimely Man," in *Twilight of the Idols* and *The Anti-Christ*, trans. R. J. Hollingdale (Penguin, 1990), 81.

27. Sigmund Freud, *The Future of an Illusion*, ed. James Strachey (Norton, 1989).

28. Edward W. Said, *Orientalism* (Penguin, 2003).

29. Daniel Dennett, *Breaking the Spell: Religion as a Natural Phenomenon* (Viking, 2006).

30. Richard Dawkins, *The Selfish Gene* (Oxford University Press, 1976).

31. Yanis Varoufakis, *Technofeudalism: What Killed Capitalism* (Bodley Head, 2024).

32. Neil McArthur, "Gods in the Machine? The Rise of Artificial Intelligence May Result in New Religions," *The Conversation*, March 15, 2023, https://theconversation.com/gods-in-the-machine-the-rise-of-artificial-intelligence-may-result-in-new-religions-201068.

33. See also Singler's discussion of the story in *Religion and Artificial Intelligence*, 43.

34. Mark Coeckelbergh, *Robot Ethics* (MIT Press, 2022).

35. Epstein, *Tech Agnostic*, 23, 116, 128.

36. Mayor, *Gods and Robots*, 1, 9–10.

37. Karla Kane, "Made, Not Born," *Palo Alto Online*, March 13, 2019, https://www.paloaltoonline.com/ae/2019/03/13/made-not-born/.

38. See for example Philip Butler, "Black Theology x Artificial Intelligence," in *The Cambridge Companion to Religion and Artificial Intelligence*, ed. Beth Singler and Fraser Watts (Cambridge University Press, 2024), 182–200.

39. In formulating this I'm inspired by Butler, "Black Theology," 198–199.

40. Deborah Netburn, "Can Religion Save Us from Artificial Intelligence?," *Los Angeles Times*, March 3, 2023, https://www.latimes.com/world-nation/story/2023-03-03/can-religion-save-us-from-artificial-inte.

41. See my own work on this topic, for example Mark Coeckelbergh, "Should We Treat Teddy Bear 2.0 as a Kantian Dog? Four Arguments for the Indirect Moral Standing of Personal Social Robots, with Implications for Thinking About Animals and Humans," *Minds and Machines* 31 (2020): 337–360.

42. Thomas Germain, "The Vatican Releases Its Own AI Ethics Handbook," *Gizmodo*, June 28, 2023, https://gizmodo.com/pope-francis-vatican-releases-ai-ethics-1850583076.

43. Given his pontifical name (Francis), emphasis on poverty, and Jesuit interest in global justice, it would be interesting to further develop links between Franciscan and Jesuit thinking, liberation theology, and contemporary political issues concerning technology.

Chapter 3

1. The eruption in 1815 had global effects and created climate abnormalities in 1816, known as the "year without a summer." Sunlight was blocked by ash and gases. Summer temperatures in Western Europe were colder than usual. In North America there was a "dry fog" that dimmed sunlight.

2. Mary Shelley, introduction to *Frankenstein*, ed. David H. Guston, Ed Finn, and Jason Scott Robert (MIT Press, 2017), 192.

3. The Western religious tradition focuses here on the child-father relationship rather than the child-mother relationship.

4. Gen. 3.

5. Rev 12:9.

6. Job 34:5, 30:20.

7. Mark 15:34.

8. See Old-New Synagogue, n.d., https://www.synagogue.cz/en/old-new-synagogue.

9. Norbert Wiener, *God & Golem, Inc.: A Comment on Certain Points Where Cybernetics Impinges on Religion* (MIT Press, 1964), 15.

10. Wiener, *God & Golem, Inc.*, 29.

11. Wiener, *God & Golem, Inc.*, 52.

12. As Yuval Harari put it recently in *Nexus* (Fern Press, 2024), waiting for a god or sorcerer to save us is dangerous since it "encourages people to abdicate responsibility and put their faith in gods and sorcerers instead," which, so he claims, are themselves human inventions (xiii). According to him, connecting to a "superhuman and infallible intelligence" is itself a fantasy (71). Seen from this perspective, waiting for a deus ex machina, whether in the form of a traditional god or a god-machine, is not only irresponsible but futile.

13. Hannah Arendt, *Eichmann in Jerusalem: A Report on the Banality of Evil* (Viking Press, 1963).

14. Wiener, *God & Golem, Inc.*, 73. See Matt. 22:21 for Jesus's words about the division of power.

15. Wiener, *God & Golem, Inc.*, 16.

16. Gen. 1.

17. Wiener, *God & Golem, Inc.*, 17.

18. When Microsoft started integrating OpenAI's ChatGPT across its applications and infrastructure in 2023, it fired its AI ethics team. Earlier, Meta dissolved its Responsible Innovation team. Cristina Criddle

and Madhumita Murgia, "Big Tech Companies Cut AI Ethics Staff, Raising Safety Concerns," *Financial Times*, March 29, 2023, https://www.ft.com/content/26372287-6fb3-457b-9e9c-f722027f36b3.

19. Gottfried Wilhelm Leibniz, *Theodicy*, ed. Austin Farrer, trans. E. M. Huggard (Yale University Press, 1952).

20. Voltaire, *Candide, or Optimism* (Penguin, 2006).

21. Gottfried Wilhelm Leibniz, *The Art of Discovery* (1685), in *Leibniz Selections*, ed. Philip P. Wiener (Charles Scribner's Sons, 1951), 51.

Chapter 4

1. For the full story of the mechanical Turk see Tom Standage, *The Turk: The Life and Times of the Famous Eighteenth-Century Chess-Playing Machine* (Walker & Company, 2002).

2. Laura Bratton, "OpenAI Just Made ChatGPT Cheaper and Twice as Fast," *Quartz*, May 13, 2024, https://qz.com/openai-chat-gpt-updates-sam-altman-magic-1851472690.

3. See for example Claude Lévi-Strauss, *The Savage Mind* (Weidenfeld and Nicolson, 1966).

4. Minsoo Kang, *Sublime Dreams of Living Machines: The Automaton in the European Imagination* (Harvard University Press, 2011), 36.

5. Sigmund Freud, "The Uncanny," in *The Uncanny*, trans. D. McLintock (Penguin Books, 2003), 121–162.

6. Masahiro Mori, "The Uncanny Valley," trans. Karl F. MacDorman and N. Kageki, *IEEE Robotics & Automation Magazine* 19, no. 2 (2012): 98–100.

7. Karl F. MacDorman and Hiroshi Ishiguro, "The Uncanny Advantage of Using Androids in Cognitive and Social Science Research," *Interaction Studies* 7, no. 3 (2006): 297–337.

8. Bruno Latour, *We Have Never Been Modern*, trans. Catherine Porter (Harvard University Press, 1993).

9. One could argue that Catholicism, mainly through the veneration of saints and the establishment of chapels and churches on sacred sites, has absorbed and integrated elements from nature religions.

10. See also Mark Coeckelbergh, "The Spirit in the Network: Models for Spirituality in a Technological Culture," *Zygon: Journal of Religion & Science* 45, no. 4 (2010): 957–978.

11. Stef Aupers, "The Revenge of the Machines: On Modernity, Digital Technology and Animism," *Asian Journal of Social Science* 30, no. 2 (2002): 203.

12. Ronald Cole-Turner, *Transhumanism and Transcendence: Christian Hope in an Age of Technological Enhancement* (Georgetown University Press, 2011).

13. Philip Hefner, *The Human Factor: Evolution, Culture, and Religion* (Fortress Press, 1993).

14. Noreen Herzfeld, *In Our Image: Artificial Intelligence and the Human Spirit* (Fortress Press, 2002), 33.

15. For some information and comments on the Lemoine case and his background see for example Andreas Matthias, "When Is an AI System Sentient? Blake Lemoine and LaMDA AI," *Daily Philosophy*, June 23, 2022, https://daily-philosophy.com/sentient-ai-lemoine-lamda/.

16. See for example Mariana Lenharo, "What Should We Do If AI Becomes Conscious? These Scientists Say It's Time for a Plan," *Nature*, December 10, 2024, https://www.nature.com/articles/d41586-024-04023-8.

17. Nicanor Perlas, *Humanity's Last Stand: The Challenge of Artificial Intelligence: A Spiritual-Scientific Response* (Temple Lodge, 2018).

18. Ray Kurzweil, *The Age of Spiritual Machines: When Computers Exceed Human Intelligence* (Viking Press, 1999).

19. Ray Kurzweil, *The Singularity Is Near: When Humans Transcend Biology* (Viking Press, 2005).

20. Erik Davis, *TechGnosis: Myth, Magic, and Mysticism in the Age of Information* (Harmony Books, 2015).

21. Fred Turner, *From Counterculture to Cyberculture* (University of Chicago Press, 2006).

22. Erik Davis, "Myth, Magic, and Mysticism in the Age of Information," *LA Review of Books*, March 29, 2015, https://lareviewofbooks.org/article/myth-magic-mysticism-age-information/.

23. Davis, *TechGnosis*, xxvi.

24. Bruce Tognazzini, "Principles, Techniques, and Ethics of Stage Magic and Their Application to Human Interface Design," *Proceedings of Interchi'93* (1993): 355–362.

25. I borrow this metaphor from Marx, who used it several times in *Capital* and wrote that "capital is dead labour which, vampire-like, lives only by sucking living labour, and lives the more, the more labour it sucks." For an analysis see Mark Neocleous, "The Political Economy of the Dead: Marx's Vampires," *History of Political Thought* 24, no. 4 (2003): 668–684.

26. See my book *New Romantic Cyborgs*.

27. See my book on this topic: Mark Coeckelbergh, *Self-Improvement* (Columbia University Press, 2022).

28. I allude here to transhumanism and related movements such as longtermism and affective altruism.

Chapter 5

1. Gabriele Trovato, Cesar Lucho, Alvaro Ramón, Renzo Ramirez, Laureano Rodriguez, and Francisco Cuellar, "The Creation of SanTO: A Robot with 'Divine' Features," in *2018 15th International Conference on Ubiquitous Robots (UR)* (IEEE, 2018), 437–442, https://ieeexplore.ieee.org/document/8442207.

2. Harumi Suzuki, "The Robot Named 'Saint': Italian Invents Robo to Assist with Prayer, Bible Quotes, and Saint of the Day," *Churchpop*, October 19, 2023, https://www.churchpop.com/the-robot-named-saint-italian-invents-robot-to-assist-with-prayer-bible-quotes-saint-of-the-day/.

3. For a brief study of prayer apps and their adoption by religions, see again Singler, *Religion and Artificial Intelligence*, 59.

4. "A Divine Connection in Your Pocket," Text with Jesus, accessed June 18, 2025, https://textwith.me/jesus/.

5. Epstein, *Tech Agnostic*, 115.

6. Epstein cites a study according to which the average American opens their phone once every four minutes. Epstein, *Tech Agnostic*, 149.

7. Gábor Ambrus, "God Is Not Like Algorithms," *AI Theology*, June 28, 2022, https://aitheology.com/author/gabor-ambrus/.

8. William James, *The Varieties of Religious Experience* (Dover, 2002), 464.

9. This also happens in human-pet relations, which are also often modeled on the master-servant model: the pet often does not understand why the master does something yet is always subject to their power.

10. Martin Buber, *I and Thou* (Continuum, 2004).

11. See also Alexandre Guilherme, "God as Thou and Prayer as Dialogue: Martin Buber's Tools for Reconciliation," *SOPHIA* 51 (2012): 365–378.

12. Emmanuel Levinas, *Totality and Infinity: An Essay on Exteriority*, trans. Alphonso Lingis (Duquesne University Press, 1969).

13. "God and Robots: Will AI Transform Religion?," posted October 23, 2021, by BBC News, YouTube, https://www.youtube.com/watch?v=JE85PTDXARM.

14. I refer here to recent calls for considering the welfare of AI. See for example Robert Long, "We Should Take AI Welfare Seriously," *Experience Machines*, November 1, 2024, https://experiencemachines.substack.com/p/we-should-take-ai-welfare-seriously.

15. Kathleen Richardson, "The Human Relationship in the Ethics of Robotics: A Call to Martin Buber's I and Thou," *AI & Society* 34 (2017): 75–82.

16. Sherry Turkle, *Alone Together: Why We Expect More from Technology and Less from Each Other* (Basic Books, 2011); *Reclaiming Conversation* (Penguin Books, 2015).

17. See for example Amanda Sharkey and Noel Sharkey, "Granny and the Robots: Ethical Issues in Robot Care for the Elderly," *Ethics and Information Technology* 14, no. 1 (2012): 27–40; and Rob Sparrow, "Robots in Aged Care: A Dystopian Future?," *AI & Society* 31, no. 4 (2016): 445–454.

18. Amanda Lagerkvist, "Yearning for a You: Faith, Doubt and Relational Expectancy in Existential Communication with Chatbots in a World on Edge," *MedieKultur: Journal of Media and Communication Research* 76 (2024): 26, 10, 25, 26.

19. See for example Mark Coeckelbergh, "Should We Treat Teddy Bear 2.0 as a Kantian Dog? Four Arguments for the Indirect Moral Standing of Personal Social Robots, with Implications for Thinking About Animals and Humans," *Minds and Machines* 31 (2020): 337–360.

20. In my previous work I have questioned what I called the "properties" approach to moral standing, for example in my book *Growing Moral Relations* (Palgrave, 2012). I have also critically discussed the claim that what happens between humans and robots is somehow not real, for example in Mark Coeckelbergh, "How to Describe and Evaluate 'Deception' Phenomena: Recasting the Metaphysics, Ethics, and Politics of ICTs in Terms of Magic and Performance and Taking a Relational and Narrative Turn," *Ethics & Information Technology* 20 (2018): 71–85. For a performative view of moral status see Mark Coeckelbergh, "How to Do Robots with Words: A Performative View of the Moral Status of Humans and Nonhumans," *Ethics and Information Technology* 25 (2023). See also Joshua Smith's performative and relational approach to robot personhood, inspired by my work and David Gunkel's, in his book *Robot Theology: Old Questions Through New Media* (Resource Publications, 2022).

21. Plato, *Phaedrus*, trans. Christopher Rowe (Penguin, 2005), 1–68, l. 275a.

22. Derrida called this Platonic tradition *phonocentric* and *logocentric*: the views that spoken language is the primary and original method of

communication and that it brings us closer to truth since words represent an external and original reality or object. Derrida questioned these views (see for example his *Of Grammatology*, trans. Gayatri Chakravorty Spivak [Johns Hopkins University Press, 2016]). I will soon say more about phonocentrism.

23. Augustine, *Confessions*, trans. Henry Chadwick (Oxford University Press, 1991), bk. 6, ch. 3.

24. Walter J. Ong, *Orality and Literacy: The Technologizing of the Word* (Routledge, 2002).

25. Marshall McLuhan, *The Gutenberg Galaxy: The Making of Typographic Man* (University of Toronto Press, 1962).

26. See Ong, *Orality and Literacy*.

27. Augustine, *Confessions*, bk. 10, ch. 13; bk. 9, ch. 6.

28. Derrida, *Of Grammatology*. See also Coeckelbergh and Gunkel, *Communicative AI*.

29. I draw an analogy here with McLuhan's famous saying, "the medium is the message." McLuhan, *Understanding Media*, chap. 1.

Chapter 6

1. For descriptions and scientific explanations of this divination, see for example William J. Broad, *The Oracle: Ancient Delphi and the Science Behind Its Lost Secrets* (Penguin, 2006).

2. Bertrand Russell, "Has Religion Made Useful Contributions to Civilization?," in *Why I Am Not a Christian*, ed. Paul Edwards (Routledge, 1996), 24.

3. Plato, *Apology*, trans. G. M. A. Grube, in *Plato, Five Dialogues*, 2nd ed., rev. John M. Cooper (Hackett Publishing, 2002), 21–44, ll. 21a, 30e.

4. Helga Nowotny, *In AI We Trust: Power, Illusion and Control of Predictive Algorithms* (Polity Press, 2021).

5. For more on AI and time, see Mark Coeckelbergh, "Time Machines: Artificial Intelligence, Process, and Narrative," *Philosophy and Technology*

34, no. 4 (2021): 1623–1638; see also Mark Coeckelbergh, *Digital Technologies, Temporality, and the Politics of Co-Existence* (Palgrave Macmillan, 2023).

6. See for example Kurzweil, *The Singularity Is Near*; Yuval, *Homo Deus*; Nick Bostrom, *Superintelligence: Paths, Dangers, Strategies* (Oxford University Press, 2014).

7. Epstein, *Tech Agnostic*, 76, 82; on the link to Russian cosmism, see Michel Eltchaninoff, *Lenin Walked on the Moon*, trans. Tina Kover (Europa Editions, 2023); on technological Gnosticism, see for example Michael Zimmerman, "Religious Motifs in Technological Posthumanism," *Western Humanities Review* 3 (2009): 67–83.

8. Bill Joy, "Why the Future Doesn't Need Us," *Wired*, April 1, 2000, https://www.wired.com/2000/04/joy-2/.

9. Dan. 7:7.

10. Dan. 9:25–26.

11. John 18:36.

12. Ernst Bloch, *The Principle of Hope*, vol. 1, trans. Neville Plaice, Stephen Plaice, and Paul Knight (MIT Press, 1995).

13. See for example Paul Boyer, *When Time Shall Be No More: Prophecy Belief in Modern American Culture* (Harvard University Press, 1992).

14. Jacques Derrida, "No Apocalypse, Not Now (Full Speed Ahead, Seven Missiles, Seven Missives)," *Diacritics* 14, no. 2 (1984): 20–31.

15. Robert M. Geraci, *Apocalyptic AI: Visions of Heaven in Robotics, Artificial Intelligence, and Virtual Reality* (Oxford University Press, 2010), 1.

16. Noreen Herzfeld, "The Eschatological Future of Artificial Intelligence," in *The Cambridge Companion to Religion and Artificial Intelligence*, ed. Beth Singler and Fraser Watts (Cambridge University Press, 2024), 148, 149.

17. Geraci, *Apocalyptic AI*, 13.

18. See in particular Kurzweil, *The Age of Spiritual Machines* and *The Singularity Is Near*, and Bostrom, *Superintelligence*.

19. Bostrom, *Superintelligence*, 2, 20.

20. Matt. 24:1–2.

21. Bostrom, *Superintelligence*, 22.

22. See Kurzweil, *The Singularity Is Near.*

23. Moravec, *Mind Children.*

24. Toby Ord, *The Precipice: Existential Risk and the Future of Humanity* (Bloomsbury Publishing, 2020); William MacAskill, *What We Owe the Future* (Basic Books, 2022). Sometimes longtermism is combined with so-called effective altruism, a movement that, influenced by utilitarianism, advocates impartially calculating the greatest good and prioritizing causes. Combined with longtermism, it aims at improving the future. Ord and MacAskill are effective altruists. For some insight into the effective altruism movement and the community around MacAskill, see, for example, Gideon Lewis-Kraus, "The Reluctant Prophet of Effective Altruism," *New Yorker*, August 8, 2022, https://www.newyorker.com/magazine/2022/08/15/the-reluctant-prophet-of-effective-altruism. For a critical take, see Émile P. Torres, "Why Effective Altruism and 'Longtermism' Are Toxic Ideologies," interview by Nathan Robinson, *Current Affairs*, May 7, 2023, https://www.currentaffairs.org/news/2023/05/why-effective-altruism-and-longtermism-are-toxic-ideologies.

25. Eliezer Yudkowsky, "Pausing AI Developments Isn't Enough. We Need to Shut It All Down," *Time*, March 29, 2023, https://time.com/6266923/ai-eliezer-yudkowsky-open-letter-not-enough/.

26. Émile P. Torres, *Were the Great Tragedies of History "Mere Ripples"? The Case Against Longtermism* (pub. by author, 2024), https://www.xriskology.com/mini-book.

27. See again Torres, "Why Effective Altruism and 'Longtermism' Are Toxic Ideologies."

28. "Pause Giant AI Experiments: An Open Letter," Future of Life Institute, March 22, 2023, https://futureoflife.org/open-letter/pause-giant-ai-experiments/.

29. Émile P. Torres, "Elon Musk, Twitter and the Future: His Long-Term Vision Is Even Weirder Than You Think," *Salon*, April 30, 2022, https://www.salon.com/2022/04/30/elon-musk-twitter-and-the-future-his-long-term-vision-is-even-weirder-than-you-think/.

30. Mark O'Connell, "Why Silicon Valley Billionaires Are Prepping for the Apocalypse in New Zealand," *The Guardian*, February 15, 2018, https://www.theguardian.com/news/2018/feb/15/why-silicon-valley-billionaires-are-prepping-for-the-apocalypse-in-new-zealand.

31. "Elon Musk 'Willing To Serve' After Donald Trump Hints at Cabinet Post for Tesla CEO," *Economic Times* (India), August 20, 2024, https://economictimes.indiatimes.com/news/international/world-news/elon-musk-willing-to-serve-after-donald-trump-hints-at-cabinet-post-for-tesla-ceo/articleshow/112641525.cms?from=mdr. As I finish this book, Musk is appointed in the Trump administration.

32. Yann LeCun (@ylecun), "How did outlandishly fringe AI doomsday cults gain so much unwarranted political influence?," X, August 16, 2024, 6:22 pm, https://x.com/ylecun/status/1824572406532542600?s=43&t=UYgAz9LVtDhX0JytmOwerg.

Chapter 7

1. Another famous religious narrative that addresses this theme is that of Siddharta Gautama, called the Buddha: a nobleman who is first protected from seeing old age, disease, and death but once he discovers these is shocked and disgusted. He becomes a monk and eventually achieves enlightenment. After many years, he finds the path toward the end of suffering.

An interesting contemporary novel and film that deals with the theme is Kazuo Ishiguro's novel *Never Let Me Go* (Faber and Faber, 2005), in which one of the characters, Tommy, runs into a field and rages against the injustice of his destiny, screaming in anger and despair. While his rage is a protest against his soon-to-arrive, too early, unnatural death (young people are raised and used for organ donations and slowly die), it can also be seen as a protest against the inevitable reality and scandal of human mortality in general (and how power dynamics decide the lives of people and how people cope with this, including rationalization of injustice and acceptance of death). The novel's title also refers to this all-too-human protest and to the attempt to find meaning and consolation in relationships.

2. Stephen Cave, *Immortality: The Quest to Live Forever and How It Drives Civilization* (Crown Publishers, 2012), 2.

3. 1 Cor. 15:26.

4. Bhagavad Gita, 2:20, translation found at https://www.holy-bhagavad-gita.org/chapter/2/verse/20.

5. Plato, *Phaedo,* trans. G. M. A. Grube, in *Plato, Five Dialogues,* 2nd ed., rev. John M. Cooper (Hackett Publishing, 2002), 93–154, l. 70c.

6. Plato, *Phaedrus,* 245c, 246c.

7. Rom. 6:4.

8. Michel Foucault, *Discipline and Punish: The Birth of the Prison,* trans. Alan Sheridan (Pantheon Books, 1977); *History of Sexuality,* vol. 1, *An Introduction,* trans. Robert Hurley (Pantheon Books, 1978).

9. Michel Foucault, "The Subject and Power," *Critical Inquiry* 8 (1982): 777–795.

10. Coeckelbergh, *Self-Improvement.*

11. Geraci, *Apocalyptic AI,* 89.

12. *Metaverse* is a term that refers to virtual worlds in which users, represented by avatars, interact and immerse themselves. It was a buzzword in 2021 when Mark Zuckerberg declared that Meta would develop a metaverse; afterward, the company redirected itself to AI.

13. Victoria Woollaston, "We'll Be Uploading Our Entire Minds to Computers by 2045 and Our Bodies Will Be Replaced by Machines Within 90 Years, Google Expert Claims," *Daily Mail,* June 19, 2013, https://www.thekurzweillibrary.com/daily-mail-well-be-uploading-our-entire-minds-to-computers-by-2045-and-our-bodies-will-be-replaced-by-machines-within-90-years-google-expert-claims.

14. See Kurzweil, *The Singularity Is Near.*

15. Woollaston, "We'll Be Uploading Our Entire Minds to Computers by 2045."

16. Sujita Sinha, "AI Afterlife: Digital Resurrection of Dead Raises Ethical Concerns," *Interesting Engineering*, May 9, 2024, https://interestingengineering.com/innovation/ai-chatbots-dead-loved-ones.

17. Areesha Lodhi, "'Never Say Goodbye': Can AI Bring the Dead Back to Life?," *Al Jazeera*, August 9, 2024, https://www.aljazeera.com/news/2024/8/9/never-say-goodbye-can-ai-bring-the-dead-back-to-life.

18. Kate Bowman, "AI Ghosts: The Promise and Perils of Digital Griefbots," Palliative Care Volunteering, January 16, 2025, https://volunteerhub.com.au/ai-ghosts-the-promise-and-perils-of-digital-griefbots/.

19. Laurence Devillers, "Will We Live On in the Form of Virtual Avatars?," *Polytechnique Insights*, February 7, 2024, https://www.polytechnique-insights.com/en/columns/digital/will-we-live-on-in-the-form-of-virtual-avatars/.

20. Aubrey de Grey and Michael Rae, *Ending Aging: The Rejuvenation Breakthroughs That Could Reverse Human Aging in Our Lifetime* (St. Martin's Press, 2007).

21. Ray Kurzweil, "A.I. Can Radically Lengthen Your Lifespan, Says Futurist Ray Kurzweil. Here's How," *Fortune*, June 25, 2024, https://fortune.com/well/article/a-i-radically-lengthen-lifespan-ray-kurzweil.

22. Cave, *Immortality*.

23. Mick Brown, "Peter Thiel: The Billionaire Tech Entrepreneur on a Mission to Cheat Death," *The Telegraph*, September 19, 2014, https://www.telegraph.co.uk/technology/11098971/Peter-Thiel-the-billionaire-tech-entrepreneur-on-a-mission-to-cheat-death.html.

24. Brown, "Peter Thiel."

25. Mark Coeckelbergh, *Human Being @ Risk: Enhancement, Technology, and the Evaluation of Vulnerability Transformations* (Springer, 2013).

26. Arguably Descartes himself had a more nuanced view of the relation between mind and body.

27. Hubert L. Dreyfus, *What Computers Still Can't Do: A Critique of Artificial Reason* (MIT Press, 1992).

28. John Searle, "Minds, Brains and Programs," *Behavioral and Brain Sciences* 3 (1980): 417–457.

29. There is an ongoing discussion about uploading and identity. See for instance the work of Susan Schneider on this topic.

30. Calvin Mercer and Tracy J. Trothen, *Religion and the Technological Future: An Introduction to Biohacking, Artificial Intelligence, and Transhumanism* (Palgrave Macmillan, 2021), 79, 166, 152. See also my summary of Christian resurrection beliefs earlier in this chapter.

31. Mercer and Trothen, *Religion and the Technological Future*, 151–152.

32. See also my remarks about this in my interpretation of Wiener in chapter 3. Here I also allude to Heidegger's famous statement, "Only a god can save us," from an interview with *Der Spiegel* magazine in 1966 that was published posthumously. For an English translation by William J. Richardson, see Thomas Sheehan, ed., *Heidegger: The Man and the Thinker* (Transaction Publishers, 1981), 45–67.

33. See especially Nietzsche, *On the Genealogy of Morals*; and *The Gay Science*.

34. See Nietzsche, *Thus Spoke Zarathustra*. Here I disagree with Stefan Lorenz Sorgner (and Max More), who argue that Nietzsche's philosophy aligns with transhumanism. According to Sorgner, Nietzsche's philosophy, in particular his idea of the Übermensch, encourages the overcoming of current human limitations. As critics have pointed out, this is a misinterpretation. Nietzsche's view is about personal transformation and overcoming: transcendence, perhaps, but a personal one and one that takes place within the limitations of the human condition. For more discussion about this topic see Stefan Lorenz Sorgner, "Nietzsche, the Overhuman, and Transhumanism," in *Nietzsche and Transhumanism: Precursor or Enemy?*, ed. Stefan Lorenz Sorgner and Yunus Tuncel (Cambridge Scholars Publishing, 2017), 14–26; and Max More, "The Overhuman in the Transhuman," *Journal of Evolution and Technology* 21, no. 1 (2010): 1–4.

Chapter 8

1. The Dartmouth Summer Research Project on Artificial Intelligence in 1956 is often considered to be the founding event of the field of artificial intelligence.

2. Tillich, *Dynamics of Faith*, 1–4.

3. See again Varoufakis, *Technofeudalism*.

4. Coeckelbergh, "The Grammars of AI."

5. Epstein, *Tech Agnostic*, 243.

6. See Blumenberg, *Work on Myth*.

7. I am influenced by Foucault here.

8. See for example Mark Coeckelbergh, "What Is Digital Humanism? A Conceptual Analysis and an Argument for a More Critical and Political Digital (Post)humanism," *Journal of Responsible Technology* 17 (2024).

9. See again Nietzsche, *Beyond Good and Evil*.

10. In an interview a few weeks before his death in 2002, Gadamer said that "Man cannot live without hope." See Jean Grondin, "Gadamer's Hope," *Renascence: Essays on Values in Literature* 56, no. 4 (2004): 287–288.

11. This is again a point against Sorgner.

Index

Publisher contact:
The MIT Press
Massachusetts Institute of Technology
77 Massachusetts Avenue, Cambridge, MA 02139
mitpress.mit.edu

EU Authorised Representative:
Easy Access System Europe, Mustamäe tee 50,
10621 Tallinn, Estonia
gpsr.requests@easproject.com

Printed by Integrated Books International,
United States of America